Subjective Probability

This book offers a concise survey of basic probability theory from a thoroughly subjective point of view whereby probability is a mode of judgment. Written by one of the greatest figures in the field of probability theory the book is both a summation and a synthesis of a lifetime of wrestling with these problems and issues.

After an introduction to basic probability theory, there are chapters on scientific hypothesis-testing, on changing your mind in response to generally uncertain observations, on expectations of the values of random variables, on de Finetti's dissolution of the so-called problem of induction, and on decision theory.

Richard Jeffrey was Emeritus Professor in the Department of Philosophy, Princeton University.

Subjective Probability

The Real Thing

RICHARD JEFFREY

Princeton University

CAMBRIDGE UNIVERSITY PRESS
Cambridge, New York, Melbourne, Madrid, Cape Town, Singapore,
São Paulo, Delhi, Dubai, Tokyo, Mexico City

Cambridge University Press
The Edinburgh Building, Cambridge CB2 8RU, UK

Published in the United States of America by Cambridge University Press, New York

www.cambridge.org
Information on this title: www.cambridge.org/9780521536684

First published 2004
Reprinted 2007 (twice)

A catalogue record for this publication is available from the British Library

Library of Congress Cataloguing in Publication Data

Jeffrey, Richard C.
Subjective probability : the real thing / Richard Jeffrey.
p. cm
Includes bibliographical references and index.
ISBN 0-521-82971-2 – ISBN 0-521-53668-5 (pbx.)
1. Probability. 2. Judgement. I. Title.
BC141.J445 2004
121'.63–dc22 2003065689

ISBN 978-0-521-82971-7 Hardback
ISBN 978-0-521-53668-4 Paperback

For Edith

Contents

Preface

Here is an account of basic probability theory from a thoroughly "subjective" point of view,[1] according to which probability is a mode of judgment. From this point of view probabilities are "in the mind"—the subject's, say, YOURS. If you say the probability of rain is 70% you are reporting that, all things considered, you would bet on rain at odds of 7:3, thinking of longer or shorter odds as giving an unmerited advantage to one side or the other.[2] A more familiar mode of judgment is flat, "dogmatic" assertion or denial, as in "It will rain" or "It will not rain". In place of this "dogmatism", the probabilistic mode of judgment offers a richer palate for depicting your state of mind, in which the colors are all the real numbers from 0 to 1. The question of the precise relationship between the two modes is a delicate one, to which I know of no satisfactory detailed answer.[3]

Chapter 1, "Probability Primer," is an introduction to basic probability theory, so conceived. The object is not so much to enunciate the formal rules of the probability calculus as to show why they must be as they are, on pain of inconsistency.

Chapter 2, "Testing Scientific Theories," brings probability theory to bear on vexed questions of scientific hypothesis-testing. It features Jon

[1] In this book double quotes are used for "as they say", where the quoted material is both used and mentioned. I use single quotes for mentioning the quoted material.

[2] This is in test cases where, over the possible range of gains and losses, your utility for income is a linear function of that income. The thought is that the same concept of probability should hold in all cases, linear or not and monetary or not.

[3] It would be a mistake to identify assertion with probability 1 and denial with probability 0, e.g., because someone who is willing to assert that it will rain need not be prepared to bet life itself on rain.

Dorling's "Bayesian" solution of Duhem's problem (and Quine's), the dreaded holism.[4]

Chapter 3, "Probability Dynamics; Collaboration," addresses the problem of changing your mind in response to generally uncertain observations of your own or your collaborators, and of packaging uncertain reports for use by others who may have different background probabilities. Conditioning on certainties is not the only way to go.

Chapter 4, "Expectation Primer," is an alternative to chapter 1 as an introduction to probability theory. It concerns your "expectations" of the values of "random variables". Hypotheses turn out to be 2-valued random variables—the values being 0 (for falsehood) and 1 (for truth). Probabilities turn out to be expectations of hypotheses.

Chapter 5, "Updating on Statistics," presents Bruno de Finetti's dissolution of the so-called "problem of induction". We begin with his often-overlooked miniature of the full-scale dissolution. At full scale, the active ingredient is seen to be "exchangeability" of probability assignments: In an exchangeable assignment, probabilities and expectations can adapt themselves to observed frequencies via conditioning on frequencies and averages "out there". Sec. **5.2,** on de Finetti's generalizations of exchangeability, is an article by other people, adapted as explained in a footnote to the section title with the permission of the authors and owners of the copyright.

Chapter 6, "Choosing," is a brief introduction to decision theory, focussed on the version floated in my *Logic of Decision* (McGraw-Hill, 1965; University of Chicago Press, 1983, 1990). It includes an analysis in terms of probability dynamics of the difference between seeing truth of one hypothesis as a probabilistic cause of truth of another hypothesis or as a mere symptom of it. In these terms, "Newcomb" problems are explained away.

Acknowledgments

Dear Comrades and Fellow Travellers in the Struggle for Bayesianism!

It has been one of the greatest joys of my life to have been given so much by so many of you. (I speak seriously, as a fond foolish old fart dying of a surfeit of Pall Malls.) It began with open-handed friendship offered

[4] According to this, scientific hypotheses "must face the tribunal of sense-experience as a corporate body." See the end of Willard van Orman Quine's much-anthologized "Two dogmas of empiricism."

to a very young beginner by people I then thought of as very old logical empiricists—Carnap in Chicago (1946–51) and Hempel in Princeton (1955–7), the sweetest guys in the world. It seems to go with the territory.

Surely the key was a sense of membership in the cadre of the righteous, standing out against the background of McKeon's running dogs among the graduate students at Chicago, and against people playing comparable parts elsewhere. I am surely missing some names, but at Chicago I especially remember the participants in Carnap's reading groups that met in his apartment when his back pain would not let him come to campus: Bill Lentz (Chicago graduate student), Ruth Barcan Marcus (post-doc, bopping down from her teaching job at Roosevelt College in the Loop), Chicago graduate student Norman Martin, Abner Shimony (visiting from Yale), Howard Stein (superscholar, all-round brain), Stan Tennenbaum (ambient mathematician) and others.

It was Carnap's booklet, *Philosophy and Logical Syntax,* that drew me into logical positivism as a 17-year-old at Boston University. As soon as they let me out of the Navy in 1946 I made a beeline for the University of Chicago to study with Carnap. And it was Hempel, at Princeton, who made me aware of de Finetti's and Savage's subjectivism, thus showing me that, after all, there was work I could be doing in probability even though Carnap had seemed to be vacuuming it all up in his 1950 and 1952 books. (I think I must have learned about Ramsey in Chicago, from Carnap.)

As to Princeton, I have Edith Kelman to thank for letting me know that, contrary to the impression I had formed at Chicago, there might be something for a fellow like me to do in a philosophy department. (I left Chicago for a job in the logical design of a digital computer at MIT, having observed that the rulers of the Chicago philosophy department regarded Carnap not as a philosopher but as—well, an engineer.) But Edith, who had been a senior at Brandeis, asssured me that the phenomenologist who taught philosophy there regarded logical positivism as a force to be reckoned with, and she assured me that philosophy departments existed where I might be accepted as a graduate student, and even get a fellowship. So it happened at Princeton. (Carnap had told me that Hempel was moving to Princeton. At that time, he himself had finally escaped Chicago for UCLA.)

This is my last book (as they say, "probably"). Like my first, it is dedicated to my beloved wife, Edith. The dedication there, "Uxori delectissimae Laviniae", was a fraudulent in-group joke, the Latin having been provided by Elizabeth Anscombe. I had never managed to learn the language. Here I come out of the closet with my love.

When the dean's review of the philosophy department's reappointment proposal gave me the heave-ho! from Stanford I applied for a visiting year at the Institute for Advanced Study in Princeton, N.J. Thanks, Bill Tait, for the suggestion; it would never have occurred to me. Gödel found the L of D interesting, and decided to sponsor me. That blew my mind.

My project was what became *The Logic of Decision* (1965). I was at the IAS for the first half of 1963–4; the other half I filled in for Hempel, teaching at Princeton University; he was on leave at the Center for Advanced Study in the Behavioral Sciences in Palo Alto. At Stanford, Donald Davidson had immediately understood what I was getting at in that project. And I seem to have had a similar sensitivity to his ideas. He was the best thing about Stanford. And the decanal heave-ho! was one of the best things that ever happened to me, for the sequel was a solution by Ethan Bolker of a mathematical problem that had been holding me up for a couple of years. Read about it in *The Logic of Decision,* especially the 2nd edition.

I had been trying to prove uniqueness of utility except for two arbitrary constants. Gödel conjectured (what Bolker turned out to have proved) that in my system there has to be a third constant, adjustable within a certain range. It was typical of Gödel that he gave me time to work the thing out on my own. We would meet every Friday for an hour before he went home for lunch. Only toward the end of my time at the IAS did he phone me, one morning, and ask if I had solved that problem yet. When I said "No" he told me he had an idea about it. I popped right over and heard it. Gödel's proof used linear algebra, which I did not understand. But, assured that Gödel had the right theorem and that some sort of algebraic proof worked, I hurried on home and worked out a clumsy proof using high school algebra. More accessible to me than Gödel's idea was Bolker's proof, using projective geometry. (In his almost finished thesis on measures on Boolean algebras Bolker had proved the theorem I needed. He added a chapter on the application to decision theory and (he asks me to say) stumbled into the honor of a citation in the same paragraph as Kurt Gödel, and I finished my book. Bliss.

My own Ph.D. dissertation was not about decision theory but about a generalization of conditioning (of "conditionalization") as a method of updating your probabilities when your new information is less than certain. This seemed to be hard for people to take in. But when I presented that idea at a conference at the University of Western Ontario in 1975 Persi Diaconis and Sandy Zabell were there and caught the spark. The result was the first serious mathematical treatment of probability kinematics (*JASA*, 1980). More bliss. (See also Carl Wagner's work, reported in chapters 2 and 3 here.)

As I addressed the problem of mining the article they had given me on partial exchangeability in volume 2 of *Studies in Inductive Logic and Probability* (1980) for recycling here, it struck me that Persi and his collaborator David Freedman could do the job far better. Feigning imminent death, I got them to agree (well, they gave me "carte blanche", so the recycling remained my job) and have obtained the necessary permission from the other interested parties, so that the second, bigger part of chapter 5 is their work. Parts of it use mathematics a bit beyond the general level of this book, but I thought it better to post warnings on those parts and leave the material here than to go through the business of petty fragmentation.

Since we met at the University of Chicago in—what? 1947?—my guru Abner Shimony has been turning that joke title into a simple description. For one thing, he proves to be a monster of intellectual integrity: After earning his Ph.D. in philosophy at Yale with his dissertation on "Dutch Book" arguments for Carnap's "strict regularity", he went back for a Ph.D. in physics at Princeton before he would present himself to the public as a philosopher of science and go on to do his "empirical metaphysics", collaborating with experimental physicists in tests of quantum theory against the Bell inequalities as reported in journals of philosophy and of experimental physics. For another, as a graduate student of physics he took time out to tutor me in Aristotle's philosophy of science while we were both graduate students at Princeton. In connection with this book, Abner has given me the highly prized *feedback* on Dorling's treatment of the Holt-Clauser issues in chapter 2, as well as other matters, here and there. Any remaining errors are his responsibility.

Over the past 30 years or so my closest and constant companion in the matters covered in this book has been Brian Skyrms. I see his traces everywhere, even where unmarked by references to publications or particular personal communications. He is my main Brother in Bayes and source of Bayesian joy.

And there are others—as, for example, Arthur Merin in far-off Munich and, close to home, Ingrid Daubechies, Our Lady of the Wavelets (alas! a closed book to me) but who does help me as a Sister in Bayes and fellow slave of LaTeX, according to my capabilities. And, in distant Bologna, my longtime friend and collaborator Maria Carla Galavotti, who has long led me in the paths of de Finetti interpretation. And I don't know where to stop: here, anyway, for now.

Except, of course, for the last and most important stop, to thank my wonderful family for their loving and effective support. Namely: The divine Edith (wife to Richard), our son Daniel and daughter Pamela, who grew unpredictably from child apples of our eyes to a union-side labor lawyer (Pam) and a computernik (Dan) and have been so generously attentive and helpful as to shame me when I think of how I treated my own ambivalently beloved parents, whose deserts were so much better than what I gave them. And I have been blessed with a literary son-in-law who is like a second son, Sean O'Connor, daddy of young Sophie Jeffrey O'Connor and slightly younger Juliet Jeffrey O'Connor, who are also incredibly kind and loving.

Let me not now start on the *Machetonim*, who are all every bit as remarkable. Maybe I can introduce you some time.

Richard Jeffrey

1
Probability Primer

Yes or no: Was there once life on Mars? We can't say. What about intelligent life? That seems most unlikely, but again, we can't really say. The simple yes-or-no framework has no place for shadings of doubt, no room to say that we see intelligent life on Mars as far less probable than life of a possibly very simple sort. Nor does it let us express exact probability judgments, if we have them. We can do better.

1.1 Bets and Probabilities

What if you were able to say exactly what odds you would give on there having been life, or intelligent life, on Mars? That would be a more nuanced form of judgment, and perhaps a more useful one. Suppose your odds were 1:9 for life, and 1:999 for intelligent life, corresponding to probabilities of 1/10 and 1/1000, respectively. (The colon is commonly used as a notation for "/", division, in giving odds—in which case it is read as "to".)

Odds m:n correspond to probability $\frac{m}{m+n}$

That means you would see no special advantage for either player in risking one dollar to gain nine in case there was once life on Mars; and it means you would see an advantage on one side or the other if those odds were shortened or lengthened. And similarly for intelligent life on Mars when the risk is a thousandth of the same ten dollars (1 cent) and the gain is 999 thousandths ($9.99).

Here is another way of saying the same thing: You would think a price of one dollar just right for a ticket worth ten if there was life on Mars and nothing if there was not, but you would think a price of only one cent right if there would have had to have been intelligent life on Mars for the ticket to be worth ten dollars. These are the two tickets:

Worth $10 if there was life on Mars.	Worth $10 if there was intelligent life on Mars.
Price: $1 Probability: 0.1	Price: 1 cent Probability: 0.001

So if you have an exact judgmental probability for truth of a hypothesis, it corresponds to your idea of the dollar value of a ticket that is worth 1 unit or nothing, depending on whether the hypothesis is true or false. (For the hypothesis of mere life on Mars the unit was $10; the price was a tenth of that.)

Of course you need not have an exact judgmental probability for life on Mars, or for intelligent life there. Still, we know that any probabilities anyone might think acceptable for those two hypotheses ought to satisfy certain rules, e.g., that the first cannot be less than the second. That is because the second hypothesis implies the first. (See the implication rule at the end of sec. **1.3** below.) In sec. **1.2** we turn to the question of what the laws of judgmental probability are, and why. Meanwhile, take some time with the following questions, as a way of getting in touch with some of your own ideas about probability. Afterward, read the discussion that follows.

Questions

1. A vigorously flipped thumbtack will land on the sidewalk. Is it reasonable for you to have a probability for the hypothesis that it will land point up?
2. An ordinary coin is to be tossed twice in the usual way. What is your probability for the head turning up both times?
   (a) 1/3, because 2 heads is one of three possibilities: 2, 1, 0 heads?
   (b) 1/4, because 2 heads is one of four possibilities: HH, HT, TH, TT?
3. There are three coins in a bag: ordinary, two-headed, and two-tailed. One is shaken out onto the table and lies head up. What should be your probability that it's the two-headed one?
   (a) 1/2, since it can only be two-headed or normal?

   (b) 2/3, because the other side could be the tail of the normal coin, or either side of the two-headed one? (Suppose the sides have microscopic labels.)

4. *It's a goy!*[1]
   (a) As you know, about 49% of recorded human births have been girls. What is your judgmental probability that the first child born after time t (say, $t =$ the beginning of the 22nd century, GMT) will be a girl?
   (b) A *goy* is defined as a girl born before t or a boy born thereafter. As you know, about 49% of recorded human births have been goys. What is your judgmental probability that the first child born in the 22nd century will be a goy?

Discussion

1. Surely it is reasonable to suspect that the geometry of the tack gives one of the outcomes a better chance of happening than the other; but if you have no clue about which of the two has the better chance, it may well be reasonable for each to have judgmental probability 1/2. Evidence about the chances might be given by statistics on tosses of similar tacks, e.g., if you learned that in 20 tosses there were 6 up's you might take the chance of up to be in the neighborhood of 30%; and whether or not you do that, you might well adopt 30% as your judgmental probability for up on the next toss.

2,3. These questions are meant to undermine the impression that judgmental probabilities can be based on analysis into cases in a way that does not already involve probabilistic judgment (e.g., the judgment that the cases are equiprobable). In either problem you can arrive at a judgmental probability by trying the experiment (or a similar one) often enough, and seeing the statistics settle down close enough to 1/2 or to 1/3 to persuade you that more trials will not reverse the indications. In each of these problems it is the finer of the two suggested analyses that happens to make more sense; but any analysis can be refined in significantly different ways, and there is no point at which the process of refinement has to stop. (Head or tail can be refined to head–facing–north or head–not–facing–north or tail.) Indeed some of these analyses

[1] This is a fanciful adaptation of Nelson Goodman's (1983, 73–74) "grue" paradox.

seem more natural or relevant than others, but that reflects the probability judgments we bring with us to the analyses.

4. *Goys and birls.* This question is meant to undermine the impression that judgmental probabilities can be based on frequencies in a way that does not already involve judgmental probabilities. Since all girls born so far have been goys, the current statistics for girls apply to goys as well: these days, about 49% of human births are goys. Then if you read probabilities off statistics in a straightforward way your probability will be 49% for each of these hypotheses:

   (1) The first child born after t will be a girl.
   (2) The first child born after t will be a goy.

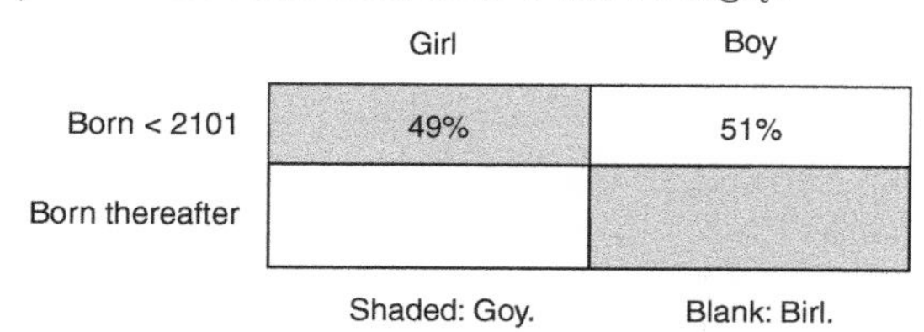

Thus $pr(1) + pr(2) = 98\%$. But it is clear that those probabilities should sum to 100%, since (2) is logically equivalent to

   (3) The first child born after t will be a boy,

and $pr(1) + pr(3) = 100\%$. To avoid this contradiction you must decide which statistics are relevant to $pr(1)$: the 49% of girls born before 2001, or the 51% of boys. And that is not a matter of statistics but of judgment—no less so because we would all make the same judgment: the 51% of boys.

1.2 Why Probabilities Are Additive

Authentic tickets of the Mars sort are hard to come by. Is the first of them really worth \$10 to me if there was life on Mars? Probably not. If the truth is not known in my lifetime, I cannot cash the ticket even if it is really a winner. But some probabilities are plausibly represented by prices, e.g., probabilities of the hypotheses about athletic contests and lotteries that people commonly bet on. And it is plausible to think that the general laws of probability ought to be the same for all hypotheses—about planets no less than about ball games. If that is so, we can justify laws of probability if we can prove all betting policies that violate them to be inconsistent. Such justifications are called "Dutch book arguments".[2] We shall give a Dutch book argument for the requirement that probabilities be additive in this sense:

> **Finite Additivity.** The probability of a hypothesis that can be true in a finite number of incompatible ways is the sum of the probabilities of those ways.

EXAMPLE 1, **Finite additivity.** The probability p of the hypothesis

(H) A sage will be elected

is $q + r + s$ if exactly three of the candidates are sages and their probabilities of winning are q, r, and s. In the following diagram, $A, B, C, D, \ldots$ are the hypotheses that the various different candidates win—the first three being the sages. The logical situation is diagrammed as follows, where the points in the big rectangle represent all the ways the election might come out, specified in minute detail, and the small rectangles represent the ways in which the winner might prove to be A, or B, or C, or D, etc.

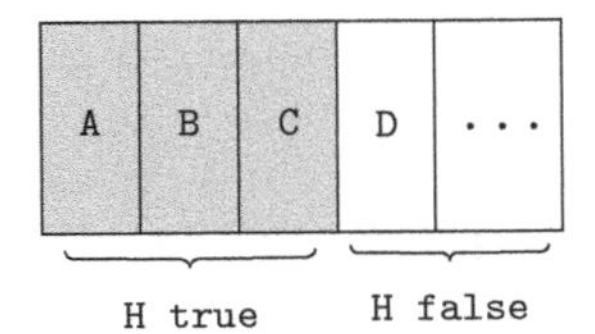

Probabilities of cases A, B, C, D,...
are q, r, s, t,..., respectively.

1.2.1 Dutch Book Argument for Finite Additivity

For definiteness we suppose that the hypothesis in question is true in three cases, as in example 1; the argument differs inessentially for other examples, with other finite numbers of cases. Now consider the following four tickets.

[2] In British racing jargon a *book* is the set of bets a bookmaker has accepted, and a book *against* someone—a "Dutch" book—is one the bookmaker will suffer a net loss on no matter how the race turns out. I follow Brian Skyrms in seeing F. P. Ramsey as holding Dutch book arguments to demonstrate actual inconsistency. See Ramsey's "Truth and Probability" in his *Philosophical Papers,* D. H. Mellor (ed.), Cambridge University Press, 1990.

`Worth $1 if H is true.`	Price $\$p$
`Worth $1 if A is true.`	Price $\$q$
`Worth $1 if B is true.`	Price $\$r$
`Worth $1 if C is true.`	Price $\$s$

Suppose I am willing to buy or sell any or all of these tickets at the stated prices. Why should p be the sum $q + r + s$? Because no matter what it is worth—\$1 or \$0—the ticket on H is worth exactly as much as the tickets on A, B, C together. (If H loses it is because A, B, C all lose; if H wins it is because exactly one of A, B, C wins.) Then if the price of the H ticket is different from the sum of the prices of the other three, I am inconsistently placing different values on one and the same contract, depending on how it is presented.

If I am inconsistent in that way, I can be fleeced by anyone who will ask me to buy the H ticket and sell or buy the other three depending on whether p is more or less than $q + r + s$. Thus, no matter whether the equation $p = q + r + s$ fails because the left—hand side is more or less than the right, a book can be made against me. That is the Dutch book argument for additivity when the number of ultimate cases under consideration is finite. The talk about being fleeced is just a way of dramatizing the inconsistency of any policy in which the dollar value of the ticket on H is anything but the sum of the values of the other three tickets: To place a different value on the three tickets on A, B, C from the value you place on the H ticket is to place different values on the same commodity bundle under two demonstrably equivalent descriptions.

When the number of cases is infinite, a Dutch book argument for additivity can still be given—provided the infinite number is not too big! It turns out that not all infinite sets are the same size.

EXAMPLE 2, **Cantor's diagonal argument.** The positive integers (I^+'s) can be counted off, just by naming them successively: 1, 2, 3, ... On the other hand, the sets of positive integers (such as {1, 2, 3, 5, 7} or the set of all even numbers, or the set of multiples of 17, or {1, 11, 101, 1001, 10001} cannot be counted off as first, second, ..., with each such set appearing as n'th in the list for some finite positive integer n. This was proved by Georg Cantor (1895) as follows. Any set of I^+'s can be represented by an endless string of plusses and minuses ("signs"), e.g., the set of even I^+'s by the string $-+-+\ldots$ in which plusses appear at the even numbered positions and minuses at the odd, the set $\{2, 3, 5, 7, \ldots\}$ of prime numbers by an endless string that begins $-++-+-+$, the set of all the I^+'s by an endless string of plusses, and the set of no I^+'s by an endless string of minuses. Cantor proved that no list of endless strings of signs can be complete. He used an amazingly simple method ("diagonalization") which, applied to any such list, yields an endless string $\bar{d}$ of signs which is not in that list. Here's how. For definiteness, suppose the first four strings in the list are the examples already given, so that the list has the general shape

$$
\begin{aligned}
s_1 &: -+-+\ldots \\
s_2 &: -++-\ldots \\
s_3 &: ++++\ldots \\
s_4 &: ----\ldots \\
&\text{etc.}
\end{aligned}
$$

Define the *diagonal* of that list as the string d consisting of the first sign in s_1, the second sign in s_2, and, in general, the n'th sign in s_n:

$$-++-\ldots$$

And Define the *antidiagonal* $\bar{d}$ of that list as the result $\bar{d}$ of reversing all the signs in the diagonal,

$$\bar{d}: +--+\ldots$$

In general, for any list $s_1, s_2, s_3, s_4, \ldots$, $\bar{d}$ cannot be any member s_n of the list, for, by definition, the n'th sign in $\bar{d}$ is different from the n'th sign of s_n, whereas if $\bar{d}$ were some s_n, those two strings would have to agree, sign by sign. Then the set of I^+'s defined by the antidiagonal of a list cannot be in that list, and therefore no list of sets of I^+'s can be complete.

> **Countability.** A countable set is defined as one whose members (if any) can be arranged in a single list, in which each member appears as the n'th item for some finite n.

Of course any finite set is countable in this sense, and some infinite sets are countable. An obvious example of a countably infinite set is the set $I^+ = \{1, 2, 3, \ldots\}$ of all positive whole numbers. A less obvious example is the set I of all the whole numbers, positive, negative, or zero: $\{\ldots, -2, -1, 0, 1, 2, \ldots\}$. The members of this set can be rearranged in a single list of the sort required in the definition of countability:

$$0, 1, -1, 2, -2, 3, -3, \ldots$$

So the set of all the whole numbers is countable. Order does not matter, as long as every member of I shows up somewhere in the list.

EXAMPLE 3, **Countable additivity.** In example 1, suppose there were an endless list of candidates, including no end of sages. If H says that a sage wins, and $A_1, A_2, \ldots$ identify the winner as the first, second, ... sage, then an extension of the law of finite additivity to countably infinite sets would be this:

> **Countable Additivity.** The probability of a hypothesis H that can be true in a countable number of incompatible ways $A_1, A_2, \ldots$ is the sum $pr(H) = pr(A_1) + pr(A_2) + \ldots$ of the probabilities of those ways.

This equation would be satisfied if the probability of one or another sage's winning were $pr(H) = 1/2$, and the probabilities of the first, second, third, etc. sage's winning were $1/4, 1/8, 1/16$, etc., decreasing by half each time.

1.2.2 Dutch Book Argument for Countable Additivity

Consider the following infinite array of tickets, where the mutually incompatible A's collectively exhaust the ways in which H can be true (as in example 3).[3]

[3] No matter that there is not enough paper in the universe for an infinity of tickets. One small ticket can save the rain forest by doing the work of all the A tickets together. This eco-ticket will say: 'For each positive whole number n, pay the bearer \$1 if A_n is true.'

`Pay the bearer $1 if H is true.`	Price $pr(H)$
`Pay the bearer $1 if` A_1 `is true.`	Price $pr(A_1)$
`Pay the bearer $1 if` A_2 `is true.`	Price $pr(A_2)$
...	...

Why should my price for the first ticket be the sum of my prices for the others? Because no matter what it is worth—$1 or $0—the first ticket is worth exactly as much as all the others together. (If H loses it is because the others all lose; if H wins it is because exactly one of the others wins.) Then if the first price is different from the sum of the others, I am inconsistently placing different values on one and the same contract, depending on how it is presented.

Failure of additivity in these cases implies inconsistency of valuations: a judgment that certain transactions are at once (1) reasonable and (2) sure to result in an overall loss. Consistency requires additivity to hold for countable sets of alternatives, finite or infinite.

1.3 Probability Logic

The simplest laws of probability are the consequences of finite additivity under this additional assumption:

> Probabilities are real numbers in the range from 0 to 1, with the endpoints reserved for certainty of falsehood and of truth, respectively.

This makes it possible to read probability laws off diagrams, much as we read ordinary logical laws off them. Let's see how that works for the ordinary ones, beginning with two surprising examples (where "iff" means *if and only if*):

De Morgan's Laws

(1) $\neg(G \wedge H) = \neg G \vee \neg H$ ("Not both true iff at least one false")

(2) $\neg(G \vee H) = \neg G \wedge \neg H$ ("Not even one true iff both false")

Here the bar, the wedge and juxtaposition stand for *not, or*, and *and.* Thus, if G and H are two hypotheses,

$G \wedge H$ (or GH) says that they are both true: G AND H

$G \vee H$ says that at least one is true: G OR H

$\neg G$ (or $-G$ or $\bar{G}$) says that G is false: NOT G

In the following diagrams for De Morgan's laws the upper and lower rows represent G and $\neg G$ and the left- and right-hand columns represent H and $\neg H$. Now if R and S are any regions, $R \wedge S$ (or "RS"), is their intersection, $R \vee S$ is their union, and $\neg R$ is the whole big rectangle except for R.

Diagrams for De Morgan's laws (1) and (2):

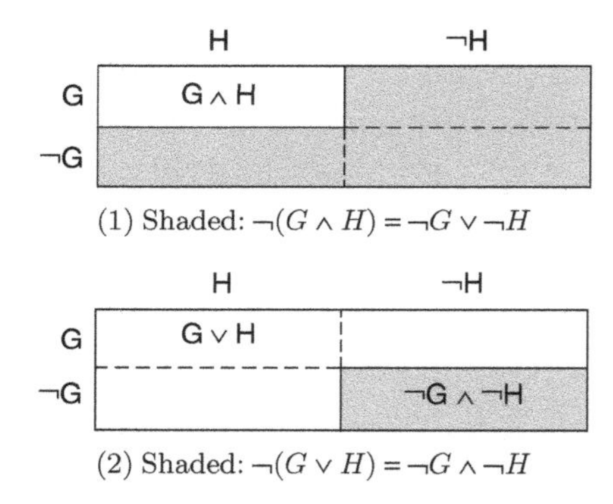

(1) Shaded: $\neg(G \wedge H) = \neg G \vee \neg H$

(2) Shaded: $\neg(G \vee H) = \neg G \wedge \neg H$

Adapting such geometrical representations to probabilistic reasoning is just a matter of thinking of the probability of a hypothesis as its region's area, assuming that the whole rectangle, $H \vee \neg H \,(= G \vee \neg G)$, has area 1. Of course the empty region, $H \wedge \neg H \,(= G \wedge \neg G)$, has area 0. It is useful to denote those two extreme regions in ways independent of any particular hypotheses H, G. Let's call them $\top$ and $\bot$:

Logical Truth. $\top = H \vee \neg H = G \vee \neg G$

Logical Falsehood. $\bot = H \wedge \neg H = G \wedge \neg G$

We can now verify some further probability laws informally, in terms of areas of diagrams.

Not: $pr(\neg H) = 1 - pr(H)$

Verification. The non-overlapping regions H and $\neg H$ exhaust the whole rectangle, which has area 1. Then $pr(H) + pr(\neg H) = 1$, so $pr(\neg H) = 1 - pr(H)$.

Or: $pr(G \vee H) = pr(G) + pr(H) - pr(G \wedge H)$

Verification. The $G \vee H$ area is the G area plus the H area, except that when you simply add $pr(G) + pr(H)$ you count the $G \wedge H$ part twice. So subtract it on the right-hand side.

The word "but"—a synonym for "and"—may be used when the conjunction may be seen as a contrast, as in "it's green but not healthy", $G \wedge \neg H$:

But Not: $pr(G \wedge \neg H) = pr(G) - pr(G \wedge H)$

Verification. The $G \wedge \bar{H}$ region is what remains of the G region after the $G \wedge H$ part is deleted.

Dyadic Analysis: $pr(G) = pr(G \wedge H) + pr(G \wedge \neg H)$

Verification. See the diagram for De Morgan (1). The G region is the union of the nonoverlapping $G \wedge H$ and $G \wedge \neg H$ regions.

In general, there is a rule of n-adic analysis for each n, e.g., for n=3:

Triadic Analysis: If H_1, H_2, H_3 *partition* $\top$,[4] then
$pr(G) = pr(G \wedge H_1) + pr(G \wedge H_2) + pr(G \wedge H_3)$.

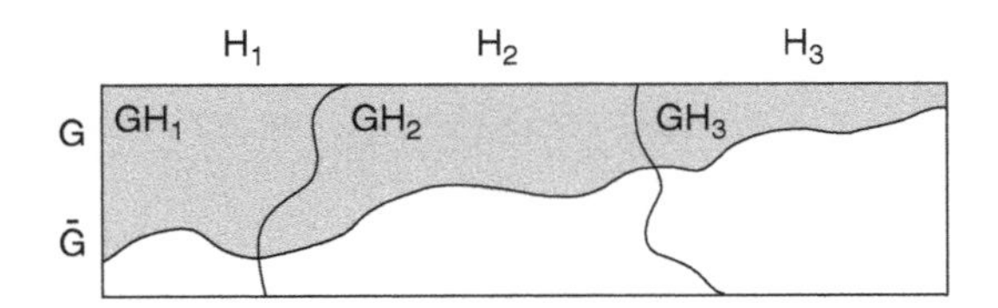

The next rule follows immediately from the fact that logically equivalent hypotheses are always represented by the same region of the diagram—in view of which we use the sign "=" of identity to indicate logical equivalence.

Equivalence: If $H = G$, then $pr(H) = pr(G)$.
(Logically equivalent hypotheses are equiprobable.)

Finally: To be implied by G, the hypothesis H must be true in every case in which G is true. Diagrammatically, this means that the G region is entirely included in the H region. In the figure below, G is represented by the small disk, and H by the large disk; H is sitting on the rim to indicate that H comprises both the annulus and the little disk. Then if

[4] This means that, as a matter of logic, the H's are mutually exclusive ($H_1 \wedge H_2 = H_1 \wedge H_3 = H_2 \wedge H_3 = \bot$) and collectively exhaustive ($H_1 \vee H_2 \vee H_3 = \top$). The equation also holds if the H's merely pr-partition $\top$ in the sense that $pr(H_i \wedge H_j) = 0$ whenever $i \neq j$ and $pr(H_1 \wedge H_2 \wedge H_3) = 1$.

G implies H, the G region can have no larger an area than the H region.

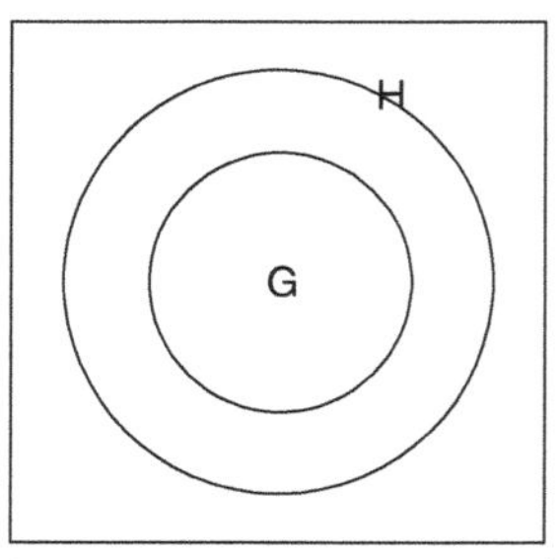

Implication: If G implies H, then $pr(G) \leq pr(H)$.

1.4 Conditional Probability

We identified your ordinary (unconditional) probability for H as the price representing your valuation of the following ticket:

Worth \$1 if H is true.	Price \$$pr(H)$

Now we identify your conditional probability for H given D as the price representing your valuation of this ticket:

Worth \$1 if $D \wedge H$ is true, Worth \$$pr(H/D)$ if D is false.	Price \$$pr(H|D)$

The old ticket represented a simple bet on H; the new one represents a conditional bet on H—a bet that is called off (the price of the ticket is refunded) in case the condition D fails. If D and H are both true, the bet is on and you win, the ticket is worth \$1. If D is true but H is false, the bet is on and you lose, the ticket is worthless. And if D is false, the bet is off, you get your \$$pr(H|D)$ back as a refund.

With that understanding we can construct a Dutch book argument for the following rule, which connects conditional and unconditional probabilities:

Product Rule: $pr(H \wedge D) = pr(H|D)pr(D)$

Dutch Book Argument for the Product Rule.[5] Imagine that you own three tickets, which you can sell at prices representing your valuations. The first is ticket (1) above. The second and third are the following two, which represent unconditional bets of \$1 on HD and of \$$pr(H|D)$ *against* D,

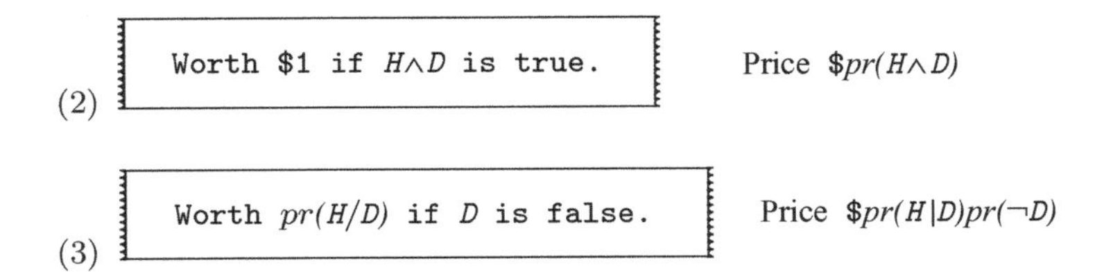

Bet (3) has a peculiar payoff: not a whole dollar, but only \$$pr(H|D)$. That is why its price is not the full \$$pr(\neg D)$ but only the fraction $pr(\neg D)$ of the \$$pr(H|D)$ that you stand to win. This payoff was chosen to equal the price of the first ticket, so that the three fit together into a neat book.

Observe that in every possible case regarding truth and falsity of H and D the tickets (2) and (3) together have the same dollar value as ticket (1). (You can verify that claim with pencil and paper.) Then there is nothing to choose between ticket (1) and tickets (2) and (3) together, and therefore it would be inconsistent to place different values on them. Thus, your price for (1) ought to equal the sum of your prices for (2) and (3):

$$pr(H|D) = pr(H \wedge D) + pr(\neg D)pr(H|D)$$

Now set $pr(\neg D) = 1 - pr(D)$, multiply through, cancel $pr(H|D)$ from both sides and solve for $pr(H \wedge D)$. The result is the product rule. To violate that rule is to place different values on the same commodity bundle in different guises: (1), or the package (2, 3).

The product rule is more familiar in a form where it is solved for the conditional probability $pr(H|G)$:

$$\textbf{Quotient Rule: } pr(H|D) = \frac{pr(H \wedge D)}{pr(D)}, \text{ provided } pr(D) > 0.$$

Graphically, the quotient rule expresses $pr(H|D)$ as the fraction of the D region that lies inside the H region. It is as if calculating $pr(H|D)$

[5] de Finetti (1937, 1980).

were a matter of trimming the whole $D \vee \neg D$ rectangle down to the D part, and using that as the new unit of area.

The quotient rule is often called the definition of conditional probability. It is not. If it were, we could never be in the position we are often in, of making a conditional judgment—say, about how a coin that may or may not be tossed will land—without attributing some particular positive value to the condition that $pr(\text{head}|\text{tossed}) = 1/2$ even though

$$\frac{pr(\text{head} \wedge \text{tossed})}{pr(\text{tossed})} = \frac{\textit{undefined}}{\textit{undefined}}.$$

Nor—perhaps, less importantly—would we be able to make judgments like the following, about a point (of area 0!) on the Earth's surface:

$$pr(\text{in western hemisphere} \mid \text{on equator}) = 1/2$$

even though

$$\frac{pr(\text{in western hemisphere} \wedge \text{on equator})}{pr(\text{on equator})} = \frac{0}{0}.$$

The quotient rule merely restates the product rule; and the product rule is no definition but an essential principle relating two distinct sorts of probability.

By applying the product rule to the terms on the right-hand sides of the analysis rules in sec. 1.3 we get the rule of[6]

Total Probability: If the D's partition $\top$ then $pr(H) = \sum_i pr(D_i)pr(H|D_i)$.[7]

EXAMPLE, A ball will be drawn blindly from urn 1 or urn 2, with odds 2:1 of being drawn from urn 2. Is black or white the more probable outcome?

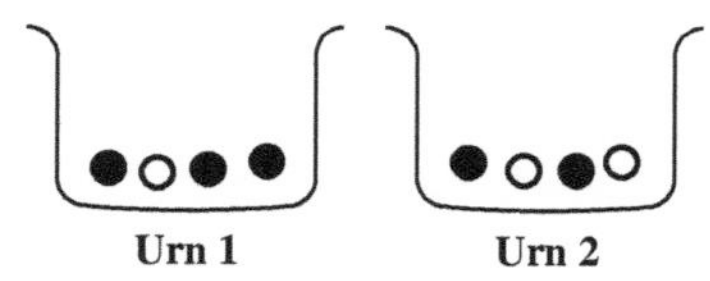

Solution. By the rule of total probability with H = black and D_i = drawn from urn i, we have $pr(H) = pr(H|D_1)P(D_1) + pr(H|D_2)P(D_2) = (\frac{3}{3} \cdot \frac{1}{3}) + (\frac{1}{2} \cdot \frac{2}{3}) = \frac{1}{4} \cdot \frac{1}{3} = \frac{7}{12} > \frac{1}{2}$: Black is the more probable outcome.

[6] Here the sequence of D's is finite or countably infinite.
[7] $= pr(D_1)pr(H|D_1) + pr(D_2)pr(H|D_2) + \ldots$

1.5 Why "|" Cannot Be a Connective

The bar in "$pr(H|D)$" is not a connective that turns pairs H, D of propositions into new, conditional propositions, H *if* D. Rather, it is as if we wrote the conditional probability of H given D as "$pr(H, D)$": The bar is a typographical variant of the comma. Thus we use "*pr*" for a function of one variable as in "$pr(D)$" and "$pr(H \wedge D)$", and also for the corresponding function of two variables as in "$pr(H|D)$". Of course the two are connected—by the product rule.

Then in fact we do not treat the bar as a statement–forming connective, "if"; but why couldn't we? What would go wrong if we did? This question was answered by David Lewis in 1976, pretty much as follows.[8] Consider the simplest special case of the rule of total probability:

$$pr(H) = pr(H|D)pr(D) + pr(H|\neg D)pr(\neg D)$$

Now if "|" is a connective and D and C are propositions, then $D|C$ is a proposition too, and we are entitled to set $H = D|C$ in the rule. Result:

(1) $\quad pr(D|C) = pr[(D|C)|D]pr(D) + pr[(D|C)|\neg D]pr(\neg D)$

So far, so good. But remember: "|" means *if*. Therefore, "$(D|C)|X$" means *If X, then if C then D*. And as we ordinarily use the word "if", this comes to the same as *If X and C, then D*:

(2) $\quad (D|C)|X = D|XC$

(Recall that the identity means the two sides represent the same region, i.e., the two sentences are logically equivalent.) Now by two applications of (2) to (1) we have

(3) $\quad pr(D|C) = pr(D|D \wedge C)pr(D) + pr(D|\neg D \wedge C)pr(\neg D)$

But as $D \wedge C$ and $\neg(D \wedge C)$ respectively imply and contradict D, we have $pr(D|D \wedge C) = 1$ and $pr(D|\neg D \wedge C) = 0$. Therefore, (3) reduces to

(4) $\quad pr(D|C) = pr(D)$

[8] For Lewis's "trivialization" result (1976), see his (1986). For subsequent developments, see the papers Hájek, *Probabilities and Conditionals*, Ellery Eells and Brian Skyrms (eds.), Cambridge University Press (1994), and Hall, *Probabilities and Conditionals*, Ellery Eells and Brian Skyrms (eds.), Cambridge University Press (1994); other papers in this book cover additional developments.

Conclusion: If "|" were a connective ("if") satisfying (2), conditional probabilities would not depend on their conditions at all. That means that "$pr(D|C)$" would be just a clumsy way of writing "$pr(D)$". And it means that $pr(D|C)$ would come to the same thing as $pr(D|\neg C)$, and as $pr(D|X)$ for any other statement X.

That is David Lewis's "trivialization result". In proving it, the only assumption needed about "if" was the eqivalence (2) of "If X, then if C then D" with "If X and C, then D".[9]

1.6 Bayes's Theorem

A well-known corollary of the product rule allows us to reverse the arguments of the conditional probability function $pr(\,|\,)$ provided we multiply the result by the ratio of the probabilities of those arguments in the original order.

Bayes's Theorem (Probabilities). $pr(H|D) = pr(D|H) \times \dfrac{pr(H)}{pr(D)}$

Proof. By the product rule, the right-hand side equals $pr(D \wedge H)/pr(D)$; by the quotient rule, so does the left-hand side.

For many purposes Bayes's theorem is more usefully applied to odds than to probabilities. In particular, suppose there are two hypotheses, H and G, to which observational data D are relevant. If we apply Bayes theorem for probabilities to the conditional *odds* between H and G, $pr(H|D)/pr(G|D)$, the unconditional probability of D cancels out:

Bayes's Theorem (Odds). $\dfrac{pr(H|D)}{pr(G|D)} = \dfrac{pr(H)}{pr(G)} \times \dfrac{pr(D|H)}{pr(D|G)}$

Terminological note. The second factor on the right-hand side of the odds form of Bayes's theorem is the "LIKELIHOOD RATIO." In these terms the odds form says:

Conditional odds = prior odds × likelihood ratio

Bayes's theorem is often stated in a form attuned to cases in which you have clear probabilities $pr(H_1), pr(H_2), \ldots,$ for mutually incompatible, collectively exhaustive hypotheses $H_1, H_2, \ldots,$ and have clear conditional probabilities $pr(D|H_1), pr(D|H_2), \ldots,$ for data D on each of them. For

[9] Note that the result does not depend on assuming that "if" means "or not"; no such fancy argument is needed in order to show that $pr(\neg A \vee B) = pr(B|A)$ only under very special conditions. (Prove it!)

a countable collection of such hypotheses we have an expression for the probability, given D, of any one of them, say, H_i:

Bayes's Theorem (Total Probabilities):[10]

$$pr(H_i|D) = \frac{pr(H_i)pr(D|H_i)}{\sum_j pr(H_j)pr(D|H_j)}$$

EXAMPLE, In the urn example (**1.4**), suppose a black ball is drawn. Was it more probably drawn from urn 1 or urn 2? Let $D =$ A black ball is drawn, $H_i =$ It came from urn i, and $pr(Hi) = \frac{1}{2}$. Here are two ways to go:

(1) Bayes's theorem for total probabilities, $pr(H_1|D) = \frac{\frac{1}{2}\cdot\frac{3}{4}}{\frac{1}{2}\cdot\frac{3}{4}+\frac{1}{2}\cdot\frac{1}{2}} = \frac{3}{5}$.

(2) Bayes's theorem for odds, $\frac{pr(H_1|D)}{pr(H_2|D)} = \frac{\frac{1}{2}}{\frac{1}{2}} \times \frac{\frac{3}{4}}{\frac{2}{4}} = \frac{3}{2}$.

These come to the same thing: Probability $3/5 =$ odds 3:2. Urn 1 is the more probable source.

1.7 Independence

Definitions

- $H_1 \wedge \ldots \wedge H_n$ is a **conjunction**. The H's are conjuncts.
- $H_1 \vee \ldots \vee H_n$ is a **disjunction**. The H's are disjuncts.
- For you, hypotheses are:

  independent iff your probability for the conjunction of any two or more is the product of your probabilities for the conjuncts;

  conditionally independent given G iff your conditional probability given G for the conjunction of any two or more is the product of your conditional probabilities given G for the conjuncts; and

  equiprobable (or conditionally equiprobable given G) iff they have the same probability for you (or the same conditional probability, given G).

EXAMPLE 1, **Rolling a fair die.** H_i means that the 1 ("ace") turns up on roll number i. These H_i are both independent and equiprobable for you: Your pr for the conjunction of any distinct n of them will be $1/6^n$.

[10] The denominator $= pr(H_1)pr(D|H_1) + pr(H_2)pr(D|H_2) + \cdots$.

EXAMPLE 2, **Coin & die.** $pr(\text{head}) = \frac{1}{2}$, $pr(1) = \frac{1}{6}$, $pr(\text{head} \wedge 1) = \frac{1}{12}$; outcomes of a toss and a roll are independent but not equiprobable.

EXAMPLE 3, **Urn of known composition.** You know it contains N balls, of which b are black, and that after a ball is drawn it is replaced and the contents of the urn mixed. $H_1, H_2, \ldots$ mean: the first, second, ... balls drawn will be black. Here, if you think nothing fishy is going on, you will regard the H's as equiprobable and independent: $pr(H_i) = b/N, pr(H_iH_j) = b^2/N^2$ if $i \neq j, pr(H_iH_jH_k) = b^3/N^3$ if $i \neq j \neq k \neq i$, and so on.

It turns out that three propositions H_1, H_2, H_3 can be independent in pairs but fail to be independent because $pr(H_1 \wedge H_2 \wedge H_3) \neq pr(H_1)$ $pr(H_2)\,pr(H_3)$.

EXAMPLE 4, **Two normal tosses of a normal coin.** If we define H_i as "head on toss $\neq i$" and D as "different results on the two tosses" we find that $pr(H_1 \wedge H_2) = pr(H_1)pr(D) = pr(H_2)pr(D) = \frac{1}{4}$ but $pr(H_1 \wedge H_2 \wedge D) = 0$.

Here is a useful fact about independence:

> (1) If n propositions are independent, so are those obtained by denying some or all of them.

To illustrate (1), think about case $n = 3$. Suppose H_1, H_2, H_3 are independent. Writing $h_i = pr(H_i)$, this means that the following four equations hold:

$$\text{(a) } pr(H_1 \wedge H_2 \wedge H_3) = h_1h_2h_3, \ \text{(b) } pr(H_1 \wedge H_2) = h_1h_2,$$
$$\text{(c) } pr(H_1 \wedge H_3) = h_1h_3, \ \text{(d) } pr(H_2 \wedge H_3) = h_2h_3$$

Here, $H_1, H_2, \neg H_3$ are also independent. Writing $\bar{h}_i = pr(\neg H_i)$, this means:

$$\text{(e) } pr(H_1 \wedge H_2 \wedge \neg H_3) = h_1h_2\bar{h}_3, \ \text{(f) } pr(H_1 \wedge H_2) = h_1h_2,$$
$$\text{(g) } pr(H_1 \wedge \neg H_3) = h_1\bar{h}_3, \ \text{(h) } pr(H_2 \wedge \neg H_3) = h_2\bar{h}_3$$

Equations $(e)-(h)$ follow from $(a)-(d)$ by the rule for "but not".

EXAMPLE 5, **(h) follows from (b).** $pr(H_2 \wedge \neg H_3) = h_2\bar{h}_3$ since, by "but not", the left-hand side $= pr(H_1) - pr(H_1 \wedge H_2)$, which, by (b), $= h_1 - h_1h_2$, and this $= h_1(1 - h_2) = h_1\bar{h}_2$.

A second useful fact follows immediately from the quotient rule:

> (2) If $pr(H_1) > 0$, then H_1 and H_2 are independent iff $pr(H_2|H_1) = pr(H_2)$.

1.8 Objective Chance

It is natural to think there is such a thing as real or objective probability ("chance", for short) in contrast to merely judgmental probability.

EXAMPLE 1, **The urn.** An urn contains 100 balls, of which an unknown number n are of the winning color—green, say. You regard the drawing as honest in the usual way. Then you think the chance of winning is n/100. If all 101 possible values of n are equiprobable in your judgment, then your prior probability for winning on any trial is 50%, and as you learn the results of more and more trials, your posterior odds between any two possible compositions of the urn will change from 1:1 to various other values, even though you are sure that the composition itself does not change.

"The chance of winning is 30%." What does that mean? In the urn example it means that $n = 30$. How can we find out whether it is true or false? In the urn example we just count the green balls and divide by the total number. But in other cases—die rolling, horse racing, etc.—there may be no process that does the "Count the green ones" job. In general, there are puzzling questions about the hypothesis that the chance of H is p that do not arise regarding the hypothesis H itself.

David Hume's skeptical answer to those questions says that chances are simply projections of ROBUST features of judgmental probabilities from our minds out into the world, whence we hear them clamoring to be let back in. That is how our knowledge that the chance of H is p guarantees that our judgmental probability for H is p : the guarantee is really a presupposition. As Hume sees it, the argument

(1) $pr(\text{the chance of } H \text{ is } p) = 1, \text{ so } pr(H) = p$

is valid because our conviction that the chance of H is p is a just as firmly felt commitment to p as our continuing judgmental probability for H.

What if we are not sure what the chance of H is, but think it may be p? Here, the relevant principle ("Homecoming") specifies the probability of H given that its chance is p—except in cases where we are antecedently sure that the chance is not p because, for some chunk $(\cdots\cdots)$ of the interval from 0 to 1, pr(the chance of H is inside the chunk) $= 0$.

> **Homecoming.** $pr(H|$ chance of H is $p) = p$ unless p is excluded as a possible value, being in the interior of an interval we are sure does not contain the chance of H :

$$0\text{–}\cdots p\cdots\text{–}1.$$

Note that when pr(the chance of H is $p) = 1$, homecoming validates argument (1). The name "Homecoming" is loaded with philosophical baggage. "Decisiveness" would be a less tendentious name, acceptable to those who see chances as objective features of the world, independent of what we may think. The condition "the chance of H is p" is decisive in that it overrides any other evidence represented in the probability function pr. But a decisive condition need not override other conditions conjoined with it to the right of the bar. In particular, it will not override the hypothesis that H is true, or that H is false. Thus, since $pr(H|H \wedge C) = 1$ when C is any condition consistent with H, we have

(2) $pr(H|H \wedge \text{the chance of } H \text{ is } .3) = 1, \text{ not } .3,$

and that is no violation of decisiveness.

On the Humean view it is ordinary conditions, making no use of the word "chance," that appear to the right of the bar in applications of the homecoming principle, e.g., conditions specifying the composition of an urn. Here, you are sure that if you knew n, your judgmental probability for the next ball's being green would be $n\%$:

(3) $pr(\text{Green next} \mid n \text{ of the hundred are green}) = n\%$

Then in example 1 you take the chance of green next to be a magnitude,

$$ch(\text{green next}) = \frac{\text{number of green balls in the urn}}{\text{total number of balls in the urn}},$$

which you can determine empirically by counting. It is the fact that for yout he ratio of green ones to all balls in the urn satisfies the decisiveness condition that identifies $n\%$ as the chance of green next, in your judgment.

On the other hand, the following two examples do not respond to that treatment.

EXAMPLE 2, **The proportion of greens drawn so far.** This won't do as your ch(green next) because it lacks the ROBUSTNESS property: On the second draw it can easily change from 0 to 1 or from 1 to 0, and until the first draw it gives the chance of green next the unrevealing form of 0/0.

EXAMPLE 3, **The loaded die.** Suppose H predicts ace on the next toss. Perhaps you are sure that if you understood the physics better, knowledge of the mass distribution in the die would determine for you a definite robust judgmental probability of ace next: You think that if you knew the physics, then you would know of an f that makes this principle true:

(4) $pr(\text{Ace next} \mid \text{The mass distribution is } M) = f(M)$

But you don't know the physics; all you know for sure is that $f(M) = \frac{1}{6}$ in case M is uniform.

When we are unlucky in this way—when there is no decisive physical parameter for us—there may still be some point in speaking of the chance of H next, i.e., of a yet-to-be-identified physical parameter that will be decisive for people in the future. In the homecoming condition we might read 'chance of H' as a place-holder for some presently unavailable description of a presently unidentified physical parameter. There is no harm in that—as long as we don't think we are already there.

In examples 1 and 2 it is clear to us what the crucial physical parameter, X, can be, and we can specify the function, f, that maps X into the chance: X can be $n/100$, in which case f is the identity function, $f(X) = X$; or X can be n, with $f(X) = n/100$. In the die example we are clear about the parameter X, but not about the function f. And in other cases we are also unclear about X, identifying it in terms of its salient effect ("blood poisoning" in the following example), while seeking clues about the cause.

EXAMPLE 4, **Blood poisoning.**

> At last, early in 1847, an accident gave Semmelweis the decisive clue for his solution of the problem. A colleague of his, Kolletschka, received a puncture wound in the finger, from the scalpel of a student with whom he was performing an autopsy, and died after an agonizing illness during which he displayed the same symptoms that Semmelweis had observed in the victims of childbed fever. Although the role of microorganisms in such infections had not yet been recognized at that time, Semmelweis realized that "cadaveric matter" which the student's scalpel had introduced into Kolletschchka's blood stream had caused his colleague's fatal illness. And the similarities between the course of Kolletschchka's disease and that of the women in his clinic led Semmelweis to the conclusion that his patients had died of the same kind of blood poisoning.[11]

But what is the common character of the X's that we had in the first two examples, and that Semmelweis lacked, and of the f's that we had in the first, but lacked in the other two? These questions belong to pragmatics, not semantics; they concern the place of these X's and f's in our processes of judgment; and their answers belong to the probability dynamics, not statics. The answers have to do with invariance of conditional judgmental probabilities as judgmental probabilities of the conditions vary. To see how that goes, let's reformulate (4) in general terms:

> (4) **Staying Home.** $pr(H|X = a) = f(a)$ if a is not in the interior of an interval that surely contains no values of X.

Here we must think of *pr* as a variable whose domain is a set of probability assignments. These may assign different values to the condition $X = a$; but, for each H and each a that is not in an excluded interval, they assign the same value, $f(a)$, to the conditional probability of H given a.

1.9 Supplements

1. (a) Find a formula for $pr(H_1 \vee H_2 \vee H_3)$ in terms of probabilities of H's and their conjunctions.
   (b) What happens in the general case, $pr(H_1 \vee H_2 \vee \ldots \vee H_n)$?

[11] Hempel (1966), p. 4.

2. *Exclusive "Or"*. The symbol "$\vee$" stands for "or" in a sense that is not meant to rule out the possibility that the hypotheses flanking it are both true: $H_1 \vee H_2 = H_1 \vee H_2 \vee (H_1 \wedge H_2)$. Let us use symbol $\underline{\vee}$ for "or" in an exclusive sense: $H_1 \underline{\vee} H_2 = (H_1 \vee H_2) \wedge \neg(H_1 \wedge H_2)$. Find formulas for
   (a) $pr(H_1 \underline{\vee} H_2)$ and
   (b) $pr(H_1 \underline{\vee} H_2 \underline{\vee} H_3)$ in terms of probabilities of H's and their conjunctions.
   (c) Does $H_1 \underline{\vee} H_2 \underline{\vee} H_3$ mean that exactly one of the three H's is true? (No.) What *does* it mean?
   (d) What does $H_1 \underline{\vee} \ldots \underline{\vee} H_n$ mean?

3. *Diagnosis.*[12] The patient has a breast mass that her physician thinks is probably benign: frequency of malignancy among women of that age, with the same symptoms, family history, and physical findings, is about 1 in 100. The physician orders a mammogram and receives the report that in the radiologist's opinion the lesion is malignant, i.e., the mammogram is positive. Based on the available statistics, the physician's probabilities for true and false positive mammogram results were as follows, and her prior probability for the patient's having cancer was 1%. What will her conditional odds on malignancy be, given the positive mammogram?

pr(row–column)	*Malignant*	*Benign*
+*mammogram*	80%	10%
−*mammogram*	20%	90%

4. *The Taxicab Problem.*[13] "A cab was involved in a hit-and-run accident at night. Two cab companies, the Green and the Blue, operate in the city. Youare given the following data:
   (a) "85% of the cabs in the city are Green, 15% are Blue.
   (b) "A witness identified the cab as Blue. The court tested the reliability of the witness under the same circumstances that existed on the night of the accident and concluded that the witness correctly identified each one of the two colors 80% of the time and failed 20% of the time. What is the probability that the cab involved in the accident was

[12] Adapted from Eddy (1982).
[13] Kahneman, Slovic, and Tversky (1982), pp. 156–158.

Blue rather than Green [i.e., conditionally on the witness's identification]?"

5. *The Device of Imaginary Results.*[14] This is meant to help you identify your prior odds—e.g., on the hypothesis "that a man is capable of extra-sensory perception, in the form of telepathy. You may imagine an experiment performed in which the man guesses 20 digits (between 0 and 9) correctly. If you feel that this would cause the probability that the man has telepathic powers to become greater than 1/2, then the [prior odds] must be assumed to be greater than 10^{-20}. [...] Similarly, if three consecutive correct guesses would leave the probability below 1/2, then the [prior odds] must be less than 10^{-3}."

   Verify these claims about the prior odds.
6. *The Rare Disease.*[15] "You are suffering from a disease that, according to your manifest symptoms, is either A or B. For a variety of demographic reasons disease A happens to be 19 times as common as B. The two diseases are equally fatal if untreated, but it is dangerous to combine the respective appropriate treatments. Your physician orders a certain test which, through the operation of a fairly well-understood causal process, always gives a unique diagnosis in such cases, and this diagnosis has been tried out on equal numbers of A and B patients and is known to be correct on 80% of those occasions. The tests report that you are suffering from disease B. Should you nevertheless opt for the treatment appropriate to A, on the supposition that the probability of your suffering from A is 19/23? Or should you opt for the treatment appropriate to B, on the supposition [...] that the probability of your suffering from B is 4/5? It is the former opinion that would be irrational for you. Indeed, on the other view, which is the one espoused in the literature, it would be a waste of time and money even to carry out the tests, since whatever their results, the base rates would still compel a more than 4/5 probability in favor of disease A. So the literature is propagating an analysis that could increase the number of deaths from a rare disease of this kind."

   Diaconis and Freedman (1981, 333–334) suggest that "the fallacy of the transposed conditional" is being committed here, i.e., confusion of the following quantities—the second of which is the

[14] From I. J. Good (1950), p. 35.
[15] L. J. Cohen (1981), see p. 329.

true positive rate of the test for B: $pr(It\ is\ B|It\ is\ diagnosed\ as\ B)$, $pr(It\ is\ diagnosed\ as\ B|It\ is\ B)$.

Use the odds form of Bayes's theorem to verify that if your prior odds on A are 19:1 and you take the true positive rate (for A, and for B) to be 80%, your posterior probability for A should be 19/23.

7. *On the Credibility of Extraordinary Stories.*[16]

"There are, broadly speaking, two different ways in which we may suppose testimony to be given. It may, in the first place, take the form of a reply to an alternative question, a question, that is, framed to be answered by yes or no. Here, of course, the possible answers are mutually contradictory, so that if one of them is not correct the other must be so:—Has A happened, yes or no?"...

"On the other hand, the testimony may take the form of a more original statement or piece of information. Instead of saying, 'Did A happen?' we may ask, 'What happened?' Here if the witness speaks the truth he must be supposed, as before, to have but one way of doing so; for the occurrence of some specific event was of course contemplated. But if he errs he has many ways of going wrong"...

   (a) In an urn with 1000 balls, one is green and the rest are red. A ball is drawn at random and seen by no one but a slightly colorblind witness, who reports that the ball was green. What is your probability that the witness was right on this occasion, if his reliability in distinguishing red from green is .9, i.e., if $pr(He\ says\ it\ is\ X|\text{It is } X) = .9$ when $X =$ Red and when $X =$ Green?

   (b) "We will now take the case in which the witness has many ways of going wrong, instead of merely one. Suppose that the balls were all numbered, from 1 to 1000, and the witness knows this fact. A ball is drawn, and he tells me that it was numbered 25, what is the probability that he is right?" In answering you are to "assume that, there being no apparent reason why he should choose one number rather than another, he will be likely to announce all the wrong ones equally often."

[16] Adapted from pp. 409 ff. of Venn (1888, 1962).

What is now your probability that the 90% reliable witness was right?

8.1. *The Three Cards.* One is red on both sides, one is black on both sides, and the other is red on one side and black on the other. One card is drawn blindly and placed on a table. If a red side is up, what is the probability that the other side is red too?

8.2. *The Three Prisoners.* An unknown two will be shot, the other freed. Alice asks the warden for the name of one other than herself who will be shot, explaining that as there must be at least one, the warden won't really be giving anything away. The warden agrees, and says that Bill will be shot. This cheers Alice up a little: Her judgmental probability for being shot is now 1/2 instead of 2/3. Show (via Bayes's theorem) that

   (a) Alice is mistaken if she thinks the warden is as likely to say "Clara" as "Bill" when he can honestly say either; but that

   (b) She is right if she thinks the warden will say "Bill" when he honestly can.

8.3. *Monty Hall.* As a contestant on a TV game show, you are invited to choose any one of three doors and receive as a prize whatever lies behind it—i.e., in one case, a car, or, in the other two, a goat. When you have chosen, the host opens one of the other two doors, behind which he knows there is a goat, and offers to let you switch your choice to the third door. Would that be wise?

9. *Causation vs. Diagnosis.*[17] "Let A be the event that before the end of next year, Peter will have installed a burglar alarm in his home. Let B denote the event that Peter's home will have been burgled before the end of next year.

   "Question: Which of the two conditional probabilities, $pr(A|B)$ or $pr(A|\neg B)$, is higher?

   "Question: Which of the two conditional probabilities, $pr(B|A)$ or $pr(B|\neg A)$, is higher?

   "A large majority of subjects (132 of 161) stated that $pr(A|B) > pr(A|\neg B)$ and that $pr(B|A) < pr(B|\neg A)$, contrary to the laws of probability."

   Substantiate this critical remark by showing that the following

[17] From p. 123 of Kahneman, Slovic, and Tversky (1982).

is a law of probability.

$$pr(A|B) > pr(A|\neg B) \text{ iff} pr(B|A) > pr(B|\neg A)$$

10. Prove the following, assuming that conditions all have probability > 0.
    (a) If A implies D then $pr(A|D) = pr(A)/pr(D)$.
    (b) If D implies A then $pr(\neg A|\neg D) = pr(\neg A)/pr(\neg D)$. (*"TJ's Lemma"*)
    (c) $pr(C|A \vee B)$ is between $pr(C|A)$ and $pr(C|B)$ if $pr(A \wedge B) = 0$.

11. *Sex Bias at Berkeley?*[18] In the fall of 1973, when 8442 men and 4321 women applied to graduate departments at U.C. Berkeley, about 44% of the men were admitted, but only about 35% of the women. It looked like sex bias against women. But when admissions were tabulated for the separate departments—as below, for the six most popular departments, which together accounted for over a third of all the applicants—there seemed to be no such bias on a department-by-department basis. And the tabulation suggested an innocent one-sentence explanation of the overall statistics. What was it?

    Hint: What do the statistics indicate about how hard the different departments (numbered A through I below) were to get into?

    A admitted 62% of 825 male applicants, 82% of 108 females.
    B admitted 63% of 560 male applicants, 68% of 25 females.
    C admitted 37% of 325 male applicants, 34% of 593 females.
    D admitted 33% of 417 male applicants, 35% of 375 females.
    E admitted 28% of 191 male applicants, 24% of 393 females.
    F admitted 6% of 373 male applicants, 7% of 341 females.

12. *The Birthday Problem.* Of twenty-three people selected at random, what is the probability that at least two have the same birthday?

$$\textit{Hint:}\quad \frac{365}{365} \times \frac{364}{365} \times \frac{363}{365} \ldots (23\ \textit{factors}) \approx .49$$

13. *How would you explain the situation to Méré?*

    "M. de Méré told me he had found a fallacy in the numbers for the following reason: With one die, the odds on throwing a

[18] Freedman, Pisani, and Purves (1978), pp. 12–15.

six in four tries are 671:625. With two dice, the odds are against throwing a double six in four tries. Yet 24 is to 36 (which is the number of pairings of the faces of two dice) as 4 is to 6 (which is the number of faces of one die). This is what made him so indignant and made him tell everybody that the propositions were inconsistent and that arithmetic was self-contradictory; but with your understanding of the principles, you will easily see that what I say is right." (Pascal to Fermat, 29 July 1654)

14. *Independence, sec.* **1.7**.
    (a) Complete the proof, begun in example 5, of the case $n = 3$ of fact (1).
    (b) Prove fact (1) in the general case. Suggestion: Use mathematical induction on the number $f = 0, 1, \ldots$ of denied H's, where $2 \leq n = t + f$.

15. *Sample Spaces, sec.* **1.3**. The diagrammatic method of this section is an agreeable representation of the set-theoretical models in which propositions are represented by sets of items called "points"—a.k.a. ("possible") "worlds" or "states" ("of nature"). If $\top = \Omega =$ the set of all such possible states in a particular set-theoretical model then $\omega \in H \subseteq \Omega$ means that the proposition H is true in possible state ω. To convert such an abstract model into a sample space it is necessary to specify the intended correspondence between actual or possible happenings and members and subsets of Ω. Not every subset of Ω need be counted as a proposition in the sample space, but the ones that do count are normally assumed to form a Boolean algebra. A Boolean "σ-algebra" is a B.A. which is closed under all countable disjunctions (and, therefore, conjunctions)—not just the finite ones. A *probability space* is a sample space with a countably additive probability assignment to the Boolean algebra $\mathcal{B}$ of propositions.

2

Testing Scientific Theories

Christian Huygens gave this account of the scientific method in the introduction to his *Treatise on Light* (1690):

> ... whereas the geometers prove their propositions by fixed and incontestable principles, here the principles are verified by the conclusions to be drawn from them; the nature of these things not allowing of this being done otherwise. It is always possible thereby to attain a degree of probability which very often is scarcely less than complete proof. To wit, when things which have been demonstrated by the principles that have been assumed correspond perfectly to the phenomena which experiment has brought under observation; especially when there are a great number of them, and further, principally, when one can imagine and foresee new phenomena which ought to follow from the hypotheses which one employs, and when one finds that therein the fact corresponds to our prevision. But if all these proofs of probability are met with in that which I propose to discuss, as it seems to me they are, this ought to be a very strong confirmation of the success of my inquiry; and it must be ill if the facts are not pretty much as I represent them.

In this chapter we interpret Huygens's methodology and extend it to the treatment of certain vexed methodological questions—especially, the Duhem-Quine problem ("holism", sec. **2.6**) and the problem of old evidence (sec. **2.7**).

2.1 Quantifications of Confirmation

The thought is that you see an episode of observation, experiment, or reasoning as confirming or infirming a hypothesis to a degree that depends on whether your probability for it increases or decreases during the episode, i.e., depending on whether your posterior probability, $new(H)$, is greater or less than your prior probability, $old(H)$. Your

Probability Increment: $new(H) - old(H)$,

is one measure of that change—positive for confirmation, negative for infirmation. Other measures are the *probability factor and the odds factor*, in both of which the turning point between confirmation and infirmation is 1 instead of 0. These are the factors by which old probabilities or odds are multiplied to get the new probabilities or odds.

$$\textbf{Probability Factor: } \pi(H) =_{df} \frac{new(H)}{old(H)}$$

By your odds on one hypothesis *against* another—say, on G against an alternative H—is meant the ratio $pr(G)/pr(H)$ of your probability of G to your probability of H; by your odds simply "on G" is meant your odds on G against $\neg G$:

$$\textbf{Odds on } G \textbf{ Against } H = \frac{pr(G)}{pr(H)}$$

$$\textbf{Odds on } G = \frac{pr(G)}{pr(\neg G)}$$

Your "odds factor"—or, better, "Bayes factor"—is the factor by which your old odds can be multiplied to get your new odds:

Bayes Factor (Odds Factor):

$$\beta(G : H) = \frac{new(G)}{new(H)} / \frac{old(G)}{old(H)} = \frac{\pi(G)}{\pi(H)}.$$

Where updating $old \mapsto new$ is by conditioning on D, your Bayes factor = your old LIKELIHOOD RATIO, $old(D|G) : old(D|H)$. Thus,

$$\beta(G : H) = \frac{old(D|G)}{old(D|H)} \text{ when } new(\ ) = old(\ |D).$$

A useful variant of the Bayes factor is its logarithm, dubbed by I. J. Good[1] the "WEIGHT OF EVIDENCE":

$$w(G:H) =_{df} log\beta(G:H)$$

As odds vary from 0 to ∞, their logarithms vary from $-\infty$ to $+\infty$. High probabilities (say, 100/101 and 1000/1001) correspond to widely spaced odds (100:1 and 1000:1); but low probabilities (say, 1/100 and 1/1000) correspond to odds (1:99 and 1:999) that are roughly as cramped as the probabilities. Going from odds to log odds moderates the spread at the high end, and treats high and low symmetrically. Thus, high odds 100:1 and 1000:1 become log odds 2 and 3, and low odds 1:100 and 1:1000 become log odds -2 and -3.

A. N. Turing was a playful, enthusiastic advocate of log odds, as his wartime code-breaking colleague I. J. Good reports:

> Turing suggested ... that it would be convenient to take over from acoustics and electrical engineering the notation of bels and decibels (db). In acoustics, for example, the bel is the logarithm to base 10 of the ratio of two intensities of sound. Similarly, if f [our β] is the factor in favour of a hypothesis, i.e., the ratio of its final to its initial odds, then we say that the hypothesis has gained $\log_{10} f$ bels or $10 \log_{10} f$ db. This may be described as the weight of evidence ... and 10 log o db may be called the plausibility corresponding to odds o. Thus ... Plausibility gained = weight of evidence.[2]
>
> The weight of evidence can be added to the initial log-odds of the hypothesis to obtain the final log-odds. If the base of the logarithms is 10, the unit of weight of evidence was called a ban by Turing (1941) who also called one tenth of a ban a deciban (abbreviated to db). I hope that one day judges, detectives, doctors and other earth-ones will routinely weigh evidence in terms of decibans because I believe the deciban is an intelligence-amplifier.[3]

[1] I. J. Good, *Probability and the Weighing of Evidence*, London, 1950, chapter 6.
[2] I. J. Good, op. cit., p. 63.
[3] I. J. Good, *Good Thinking*, Minneapolis, 1983, p. 132. Good means "intelligence" in the sense of "information": Think MI (military intelligence), not IQ.

> The reason for the name ban was that tens of thousands of sheets were printed in the town of Banbury on which weights of evidence were entered in decibans for carrying out an important classified process called Banburismus. A deciban or half-deciban is about the smallest change in weight of evidence that is directly perceptible to human intuition. [...] The main application of the deciban was ... for discriminating between hypotheses, just as in clinical trials or in medical diagnosis. [Good (1979), p. 394][4]

Playfulness $\neq$ frivolity. Good is making a serious proposal, for which he claims support by extensive testing in the early 1940's at Bletchley Park, where the "Banburismus" language for hypothesis-testing was used in breaking each day's German naval "Enigma" code during World War II.[5]

2.2 Observation and Sufficiency

In this chapter we confine ourselves to the most familiar way of updating probabilities, that is, conditioning on a statement when an observation assures us of its truth.[6]

EXAMPLE 1, **Huygens on light.** Let H be the conjunction of Huygens's hypotheses about light and let C represent one of the "new phenomena which ought to follow from the hypotheses which one employs".

[4] I. J. Good, "A. M. Turing's Statistical Work in World War II", *Biometrika* **66** (1979) 393–396, p. 394.

[5] See Andrew Hodges, *Alan Turing, the Enigma*, New York, 1983. The dead hand of the British Official Secrets Act is loosening; see the A. N. Turing home page, http:\\www.turing.org.uk\turing\, maintained by Andrew Hodges.

[6] In chapter 3 we consider updating by a generalization of conditioning, using statements that an observation makes merely probable to various degrees.

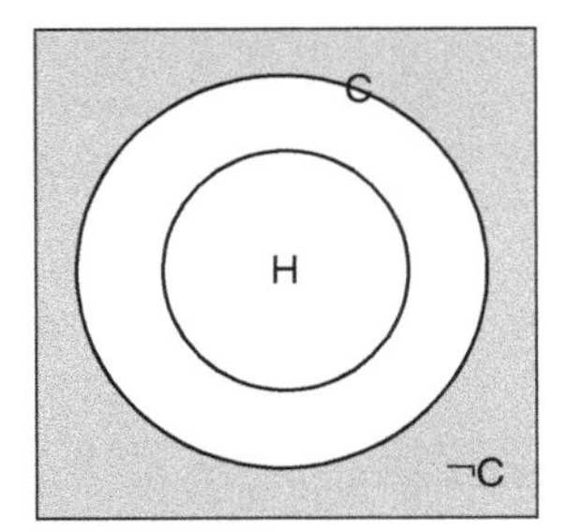

C (unshaded) is known to follow from *H*.

If we know that C follows from H, and we can discover by observation whether C is true or false, then we have the means to test H—more or less conclusively, depending on whether we find that C is false or true. If C proves false (shaded region), H is refuted decisively, for $H \wedge \neg C = \bot$. On the other hand, if C proves true, H's probability changes from

$$old(H) = \frac{\text{area of the } H \text{ circle}}{\text{area of the } \top \text{ square}} = \frac{old(H)}{1}$$

to

$$new(H) = old(H|C) = \frac{\text{area of the } H \text{ circle}}{\text{area of the } C \text{ circle}} = \frac{old(H)}{old(C)}$$

so that the probability factor is

$$\pi(H) = \frac{1}{old(C)}$$

It is the antecedently least probable conclusions whose unexpected verification raises H's probability the most.[7] The corollary to Bayes's theorem in section **1.6** generalizes this result to cases where $pr(C|H) < 1$.

Before going on, we note a small point that will loom larger in chapter **3**: In case of a positive observational result it may be inappropriate to update by conditioning on the simple statement C, for the observation

[7] "More danger, more honor": George Pólya, *Patterns of Plausible Inference*, 2nd ed., Princeton, 1968, vol. 2, p. 126.

may provide further information which overflows that package. In example 1 this problem arises when the observation warrants a report of form $C \wedge E$, which would be represented by a subregion of the C circle, with $old(C \wedge E) < old(C)$. Here (unless there is further overflow) the proposition to condition on is $C \wedge E$, not C. But things can get even trickier, as in this homely jellybean example:

EXAMPLE 2, **The green bean.**[8]

H, This bean is lime-flavored. C, This bean is green.

You are drawing a bean from a bag in which you know that half of the beans are green, all the lime-flavored ones are green, and the green ones are equally divided between lime and mint flavors. So before looking at the bean or tasting it, your probabilities are these: $old(C) = 1/2 = old(H|C); old(H) = 1/4$. But although $new(C) = 1$, your probability $new(H)$ for lime can drop below $old(H) = 1/4$ instead of rising to $old(H|C) = 1/2$ in case you yourself are the observer—for instance, if, when you see that the bean is green, you also get a whiff of mint, or also see that the bean is of a special shade of green that you have found to be associated with the mint-flavored ones. And here there need be no further proposition E you can articulate that would make $C \wedge E$ the thing to condition on, for you may have no words for the telltale shade of green, or for the while of mint. The difficulty will be especially bothersome as an obstacle to collaboration with others who would like to update their own prior probabilities by conditioning on your findings. (In chapter 3 we will see what to do about it.)

2.3 Leverrier on Neptune

We now turn to a methodological story more recent than Huygens's:[9]

> On the basis of Newton's theory, the astronomers tried to compute the motions of [...] the planet Uranus; the differences between theory and observation seemed to exceed the admissible limits of error. Some astronomers suspected that these deviations might be due to the attraction of a planet revolving beyond Uranus's orbit, and the

[8] Brian Skyrms, 'Dynamic Coherence and Probability Kinematics', *Philosophy of Science* **54** (1987) 1–20.
[9] George Pólya, *op. cit.*, pp. 130–132.

> French astronomer Leverrier investigated this conjecture more thoroughly than his colleagues. Examining the various explanations proposed, he found that there was just one that could account for the observed irregularities in Uranus's motion: the existence of an extra-Uranian planet [sc., Neptune]. He tried to compute the orbit of such a hypothetical planet from the irregularities of Uranus. Finally Leverrier succeeded in assigning a definite position in the sky to the hypothetical planet [say, with a 1 degree margin of error]. He wrote about it to another astronomer whose observatory was the best equipped to examine that portion of the sky. The letter arrived on the 23rd of September 1846 and in the evening of the same day a new planet was found within one degree of the spot indicated by Leverrier. It was a large ultra-Uranian planet that had approximately the mass and orbit predicted by Leverrier.

We treated Huygens's conclusion as a strict deductive consequence of his principles. But Pólya made the more realistic assumption that Leverrier's prediction C (a bright spot near a certain point in the sky at a certain time) was highly probable but not 100%, given his H (namely, Newton's laws and observational data about Uranus). So $old(C|H) \approx 1$. And presumably the rigidity condition was satisfied so that $new(C|H) \approx 1$, too. Then verification of C would have raised H's probability by a factor $\approx 1/old(C)$, which is large if the prior probability $old(C)$ of Leverrier's prediction was ≈ 0.

Pólya offers a reason for regarding $1/old(C)$ as at least 180, and perhaps as much as 13131: The accuracy of Leverrier's prediction proved to be better than 1 degree, and the probability of a randomly selected point on a circle or on a sphere being closer than 1 degree to a previously specified point is 1/180 for a circle, and about 1/13131 for a sphere. Favoring the circle is the fact that the orbits of all known planets lie in the same plane ("the ecliptic"). Then the great circle cut out by that plane gets the lion's share of probability. Thus, if $old(C)$ is half of 1%, H's probability factor will be about 200.

2.4 Dorling on the Duhem Problem

Skeptical conclusions about the possibility of scientific hypothesis-testing are often drawn from the presumed arbitrariness of answers to the question of which to give up—theory, or auxiliary hypothesis—when they jointly contradict empirical data. The problem, addressed by Duhem in the first years of the 20th century, was agitated by Quine in mid-century. As drawn by some of Quine's readers, the conclusion depends on his assumption that aside from our genetical and cultural heritage, deductive logic is all we've got to go on when it comes to theory testing. That would leave things pretty much as Descartes saw them, just before the mid-17th century emergence in the hands of Fermat, Pascal, Huygens and others of the probabilistic ("Bayesian") methodology that Jon Dorling has brought to bear on various episodes in the history of science.

The conclusion is one that scientists themselves generally dismiss, thinking they have good reason to evaluate the effects of evidence as they do, but regarding formulation and justification of such reasons as someone else's job—the methodologist's. Here is an introduction to Dorling's work on that job, using extracts from his important but still unpublished 1982 paper[10]—which is reproduced in the web page `http://www.princeton.edu/~bayesway` with Dorling's permission.

The material is presented here in terms of odds factors. Assuming rigidity relative to D, the odds factor for a theory T against an alternative theory S that is due to learning that D is true will be the left-hand side of the following equation, the right-hand side of which is called "the likelihood ratio":

Bayes Factor = Likelihood Ratio:

$$\frac{old(T|D)/old(S|D)}{old(T)/old(S)} = \frac{old(D|T)}{old(D|S)}$$

The empirical result D is not generally deducible or refutable by T alone, or by S alone, but in interesting cases of scientific hypothesis testing D is deducible or refutable on the basis of the theory and an

[10] Jon Dorling, "Bayesian personalism, the methodology of research programmes, and Duhem's problem", *Studies in History and Philosophy of Science* **10** (1979) 177–187. More along the same lines: Michael Redhead, "A Bayesian reconstruction of the methodology of scientific research programmes," *Studies in History and Philosophy of Science* **11** (1980) 341–347. Dorling's unpublished paper from which excerpts appear here in sec. **2.4.1–2.4.4** is "Further illustrations of the Bayesian solution of Duhem's problem" (29 pp., photocopied, 1982). References here ("sec. **4**", etc.) are to the numbered sections of that paper. Dorling's work is also discussed in Colin Howson and Peter Urbach, *Scientific Reasoning: the Bayesian approach* La Salle, Illinois, Open Court, 2nd ed., 1993.

auxiliary hypothesis A (e.g., the hypothesis that the equipment is in good working order). To simplify the analysis, Dorling makes an assumption that can generally be justified by appropriate formulation of the auxiliary hypothesis:

Prior Independence

$$old(A \wedge T) = old(A)old(T),$$
$$old(A \wedge S) = old(A)old(S).$$

Generally speaking, S is not simply the denial of T, but a definite scientific theory in its own right, or a disjunction of such theories, all of which agree on the phenomenon of interest, so that, as an explanation of that phenomenon, S is a rival to T. In any case Dorling uses the independence assumption to expand the right-hand side of the Bayes Factor = Likelihood Ratio equation:

$$\beta(T:S) = \frac{old(D|T \wedge A)old(A) + old(D|T \wedge \neg A)old(\neg A)}{old(D|S \wedge A)old(A) + old(D|S \wedge \neg A)old(\neg A)}$$

To study the effect of D on A, he also expands $\beta(A := A)$ with respect to T (and similarly with respect to S, although we do not show that here):

$$\beta(A:\neg A) = \frac{old(D|A \wedge T)old(T) + old(D|A \wedge \neg T)old(\neg T)}{old(D|\neg A \wedge T)old(T) + old(D|\neg A \wedge \neg T)old(\neg T)}$$

2.4.1 Einstein/Newton, 1919

In these terms Dorling analyzes two famous tests that were duplicated, with apparatus differing in seemingly unimportant ways, with conflicting results: one of the duplicates confirmed T against S, the other confirmed S against T. But in each case the scientific experts took the experiments to clearly confirm one of the rivals against the other. Dorling explains why the experts were right:

> In the solar eclipse experiments of 1919, the telescopic observations were made in two locations, but only in one location was the weather good enough to obtain easily interpretable results. Here, at Sobral, there were two telescopes: one, the one we hear about, confirmed Einstein; the other, in fact the slightly larger one, confirmed Newton. Conclusion: Einstein was vindicated, and the results with the larger telescope were rejected. (Dorling, sec. 4)

Notation
T: Einstein: light-bending effect of the sun
S: Newton: no light-bending effect of the sun
A: Both telescopes are working correctly
D: The conflicting data from both telescopes

In the odds factor $\beta(T:S)$ above, $old(D|T \wedge A) = old(D|S \wedge A) = 0$ since if both telescopes were working correctly they would not have given contradictory results. Then the first terms of the sums in numerator and denominator vanish, so that the factors $old(\neg T)$ cancel, and we have

$$\beta(T:S) = \frac{old(D|T \wedge \neg A)}{old(D|S \wedge \neg A)}$$

> Now the experimenters argued that one way in which A might easily be false was if the mirror of one or the other of the telescopes had distorted in the heat, and this was much more likely to have happened with the larger mirror belonging to the telescope which confirmed S than with the smaller mirror belonging to the telescope which confirmed T. Now the effect of mirror distortion of the kind envisaged would be to shift the recorded images of the stars from the positions predicted by T to or beyond those predicted by S. Hence $old(D|T \wedge \neg A)$ was regarded as having an appreciable value, while, since it was very hard to think of any similar effect which could have shifted the positions of the stars in the other telescope from those predicted by S to those predicted by $T, old(D|S \wedge \neg A)$ was regarded as negligibly small, hence the result as overall a decisive confirmation of T and refutation of S.

Thus the Bayes factor $\beta(T:S)$ is very much greater than 1.

2.4.2 Bell's Inequalities: Holt/Clauser

> Holt's experiments were conducted first and confirmed the predictions of the local hidden variable theories and refuted those of the quantum theory. Clauser examined Holt's apparatus and could find nothing wrong with it, and obtained the same results as Holt with Holt's apparatus. Holt refrained

> from publishing his results, but Clauser published his, and they were rightly taken as excellent evidence for the quantum theory and against hidden-variable theories. (Dorling, sec. 4)

Notation
T: Quantum theory
S: Disjunction of local hidden variable theories
A: Holt's setup is reliable[11] enough to distinguish T from S
D: The specific correlations predicted by T and contradicted by S are not detected by Holt's setup

The characterization of D yields the first two of the following equations. In conjunction with the characterization of A it also yields $old(D|T \wedge \neg A) = 1$, for if A is false, Holt's apparatus was not sensitive enough to detect the correlations that would have been present according to T; and it yields $old(D|S \wedge \neg A) = 1$ because of the wild improbability of the apparatus "hallucinating" those specific correlations.[12]

$$old(D|T \wedge A) = 0,$$
$$old(D|S \wedge A) = old(D|T \wedge \neg A) = old(D|S \wedge \neg A) = 1$$

Substituting these values, we have

$$\beta(T, S) = old(\neg A)$$

Then with a prior probability of 4/5 for adequate sensitivity of Holt's apparatus, the odds between quantum theory and the local hidden variable theories shift strongly in favor of the latter, e.g., with prior odds 45:55 between T and S, the posterior odds are only 9:55, a 14% probability for T.

Now why didn't Holt publish his result?

[11] This means: sufficiently sensitive and free from systematic error. Holt's setup proved to incorporate systematic error arising from tension in tubes for mercury vapor, which made the glass optically sensitive so as to rotate polarized light. Thanks to Abner Shimony for clarifying this. See also his supplement **6** in sec. **2.6.**

[12] Recall Venn on the credibility of extraordinary stories: Supplement **7** in sec. **1.11.**

Because the experimental result undermined confidence in his apparatus. Setting $\neg T = S$ in (2) because T and S were the only theories given any credence as explanations of the results, and making the same substitutions as in (4), we have

$$\beta(A : \neg A) = old(S)$$

so the odds on A fall from 4:1 to 2.2:1; the probability of A falls from 80% to 69%. Holt is not prepared to publish with better than a 30% chance that his apparatus could have missed actual quantum mechanical correlations; *the swing to S depends too much on a prior confidence in the experimental setup that is undermined by the same thing that caused the swing.*

Why did Clauser publish?

Notation

T: Quantum theory

S: Disjunction of local hidden variable theories

C: Clauser's setup is reliable enough

E: The specific correlations predicted by T and contradicted by S are detected by Clauser's setup

Suppose that $old(C) = .5$. At this point, although $old(A)$ has fallen by 11%, both experimenters still trust Holt's well-tried setup better than Clauser's. Suppose Clauser's initial results E indicate presence of the quantum mechanical correlations pretty strongly, but still with a 1% chance of error. Then E strongly favors T over S:[13]

$$\beta(T, S) = \frac{old(E|T \wedge C)old(C) + old(E|T \wedge \neg C)old(\neg C)}{old(E|S \wedge C)old(C) + old(E|S \wedge \neg C)old(\neg C)} = 50.5$$

Starting from the low 9:55 to which T's odds fell after Holt's experiment, odds after Clauser's experiment will be 909:110, an 89% probability for T. The result E boosts confidence in Clauser's apparatus by a factor of

$$\beta(C : \neg C) = \frac{old(E|C \wedge T)old(T) + old(E|C \wedge S)old(S)}{old(E|\neg C \wedge T)old(T) + old(E|\neg C \wedge S)old(S)} = 15$$

[13] Numerically: $\frac{1 \times .5 + .01 \times .5}{.01 \times .5 + .01 \times .5} = 50.5$.

This raises the initially even odds on C to 15:1, raises the probability from 50% to 94%, and lowers the 50% probability of the effect's being due to chance down to 6 or 7 percent.

2.4.3 Laplace/Adams

Finally, note one more class of cases: A theory T remains highly probable although (with auxiliary hypothesis A) it is incompatible with subsequently observed data D. With $S = \neg T$ in the formulas for β in **2.4** and with $old(D|T \wedge A) = 0$ (so that the first terms in the numerators vanish), and writing

$$t = \frac{old(D|T \wedge \neg A)}{old(D|\neg T \wedge \neg A)}, \qquad s = \frac{old(D|\neg T \wedge A)}{old(D|\neg T \wedge \neg A)},$$

for two of the likelihood ratios, it is straightforward to show that

$$\beta(T : \neg T) = \frac{t}{1 + (s \times \text{old odds on} A)},$$

$$\beta(A : \neg A) = \frac{s}{1 + (t \times \text{old odds on} T)},$$

$$\beta(T : A) = \frac{t}{s} \times \frac{old(\neg A)}{old(\neg T)}.$$

These formulas apply to

> a famous episode from the history of astronomy which clearly illustrated striking asymmetries in "normal" scientists' reactions to confirmation and refutation. This particular historical case furnished an almost perfect controlled experiment from a philosophical point of view, because owing to a mathematical error of Laplace, later corrected by Adams, the same observational data were first seen by scientists as confirmatory and later as disconfirmatory of the orthodox theory. Yet their reactions were strikingly asymmetric: what was initially seen as a great triumph and of striking evidential weight in favour of the Newtonian theory, was later, when it had to be re-analyzed as disconfirmatory after the discovery of Laplace's mathematical oversight, viewed merely as a minor embarrassment and of negligible evidential weight against the Newtonian theory. Scientists reacted in the "refutation" situation by making a hidden auxiliary hypothesis, which had previously been

> considered plausible, bear the brunt of the refutation, or, if you like, by introducing that hypothesis's negation as an apparently ad hoc face-saving auxiliary hypothesis. (Dorling, sec. 1)

Notation
T: The theory, Newtonian celestial mechanics
A: The hypothesis that disturbances (tidal friction, etc.) make a negligible contribution to
D: The observed secular acceleration of the moon

Dorling argues on scientific and historical grounds for approximate numerical values

$$t = 1, \qquad s = \frac{1}{50}.$$

The thought is that $t = 1$ because with A false, truth or falsity of T is irrelevant to D, and $t = 50s$ because in plausible partitions of $\neg T$ into rival theories predicting lunar accelerations, $old(R|\neg T) = 1/50$ where R is the disjunction of rivals not embarrassed by D. Then for a theorist whose odds are 3:2 on A and 9:1 on T (probabilities 60% for A and 90% for T),

$$\beta(T : \neg T) = \frac{100}{103}$$
$$\beta(A : \neg A) = \frac{1}{500}$$
$$\beta(T : A) = 200$$

Thus the prior odds 900:100 on T barely decrease, to 900:103; the new probability of T, 900/1003, agrees with the original 90% to two decimal places. But odds on the auxiliary hypothesis A drop sharply, from prior 3:2 to posterior 3/1000, i.e., the probability of A drops from 60% to about three tenths of 1%; odds on the theory against the auxiliary hypothesis increase by a factor of 200, from a prior value of 3:2 to a posterior value of 300:1.

2.4.4 Dorling's Conclusions

> Until recently there was no adequate theory available of how scientists should change their beliefs in the light of

evidence. Standard logic is obviously inadequate to solve this problem unless supplemented by an account of the logical relations between degrees of belief which fall short of certainty. Subjective probability theory provides such an account and is the simplest such account that we possess. When applied to the celebrated Duhem (or Duhem-Quine) problem and to the related problem of the use of ad hoc, or supposedly ad hoc, hypotheses in science, it yields an elegant solution. This solution has all the properties which scientists and philosophers might hope for. It provides standards for the validity of informal inductive reasoning comparable to those which traditional logic has provided for the validity of informal deductive reasoning. These standards can be provided with a rationale and justification quite independent of any appeal to the actual practice of scientists, or to the past success of such practices.[14] Nevertheless they seem fully in line with the intuitions of scientists in simple cases and with the intuitions of the most experienced and most successful scientists in trickier and more complex cases. The Bayesian analysis indeed vindicates the rationality of experienced scientists' reactions in many cases where those reactions were superficially paradoxical and where the relevant scientists themselves must have puzzled over the correctness of their own intuitive reactions to the evidence. It is clear that in many such complex situations many less experienced commentators and critics have sometimes drawn incorrect conclusions and have mistakenly attributed the correct conclusions of the experts to scientific dogmatism. Recent philosophical and sociological commentators have sometimes generalized this mistaken reaction into a full-scale attack on the rationality of men of science, and as a result have mistakenly looked

[14] Here a long footnote explains the Putnam-Lewis Dutch book argument for conditioning. Putnam stated the result, or, anyway, a special case, in a 1963 *Voice of America* Broadcast, 'Probability and Confirmation', reprinted in his *Mathematics, Matter and Method*, Cambridge University Press, 1975, 293–304. Paul Teller, 'Conditionalization and observation', *Synthese* **26** (1973) 218–258, reports a general argument to that effect which Lewis took to have been what Putnam had in mind.

> for purely sociological explanations for many changes scientists' beliefs, or the absence of such changes, which were in fact, as we now see, rationally de rigeur. (Dorling, sec. **5**)

2.5 Old News Explained

In sec. **2.2** we analyzed the case Huygens identified as the principal one in his *Treatise on Light:* A prediction C long known to follow from a hypothesis H is now found to be true. Here, if the rigidity condition is satisfied, $new(H) = old(H|C)$, so that the probability factor is $\pi(H) = 1/old(C)$.

But what if some long-familiar phenomenon C, a phenomenon for which $old(C) = 1$, is newly found to follow from H in conjunction with familiar background information B about the solar system, and thus to be explained by $H \wedge B$? Here, if we were to update by conditioning on C, the probability factor would be 1 and $new(H)$ would be the same as $old(H)$. No confirmation here.[15]

Wrong, says Daniel Garber:[16] The information prompting the update is not that C is true, but that $H \wedge B$ implies C. To condition on that news Garber proposes to enlarge the domain of the probability function *old* by adding to it the hypothesis that C follows from $H \wedge B$ together with all further truth-functional compounds of that new hypothesis with the *old* domain. Using some extension *old** of *old* to the enlarged domain, we might then have $new(H) = old^*(H|H \wedge B \text{ implies } C)$. That is an attractive approach to the problem, if it can be made to work.[17]

The alternative approach that we now illustrate Defines the *new* assignment on the same domain that *old* had. It analyzes the $old \mapsto new$ transition by embedding it in a larger process, $ur \overset{obs}{\mapsto} old \overset{expl}{\mapsto} new$, in which *ur* represents an original state of ignorance of C's truth and its logical relationship to H:

[15] This is what Clark Glymour has dubbed "the paradox of old evidence": see his *Theory and Evidence*, Princeton University Press, 1980.

[16] See his "Old evidence and logical omniscience in Bayesian confirmation theory" in *Testing Scientific Theories*, John Earman (ed.), University of Minnesota Press, 1983. For further discussion of this proposal see "Bayesianism with a human face" in my *Probability and the Art of Judgment*, Cambridge University Press, 1992.

[17] For a critique, see chapter 5 of John Earman's *Bayes or Bust?*, Cambridge, Mass., MIT Press, 1992.

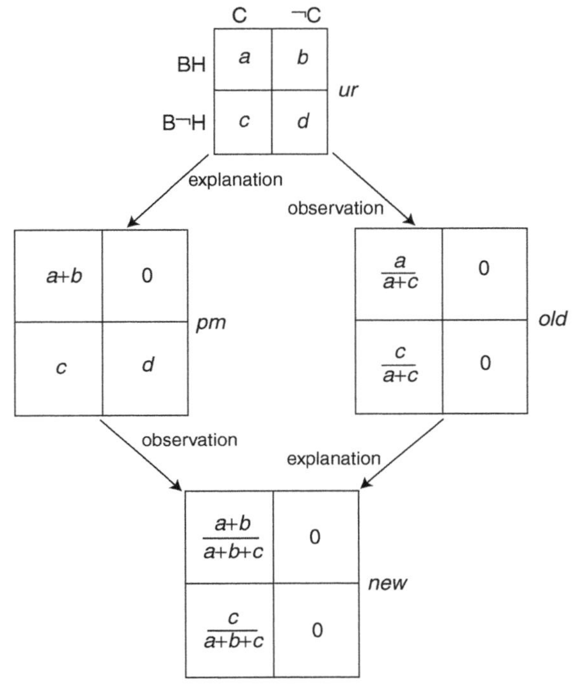

Confirmation by New Explanation

EXAMPLE 1, **The perihelion of Mercury.**
Notation.
H: GTR applied to the Sun-Mercury system
C: Advance of 43 seconds of arc per century[18]

In 1915 Einstein presented a report to the Prussian Academy of Sciences explaining the then-known advance of approximately 43″ per century in the perihelion of Mercury in terms of his ("GTR") field equations for gravitation. An advance of 38″ per century had been reported by Leverrier in 1859, due to "some unknown action on which no light has been thrown".[19]

[18] Applied to various Sun-planet systems, the GTR says that all planets' perihelions advance, but that Mercury is the only one for which that advance should be observable. This figure for Mercury has remained good since its publication in 1882 by Simon Newcomb.
[19] Pais, *"Subtle is the Lord . . .", the Science and Life of Albert Einstein*, Oxford University Press, 1982, p. 254.

> The only way to explain the effect would be ([Leverrier] noted) to increase the mass of Venus by at least 10 percent, an inadmissible modification. He strongly doubted that an intramercurial planet ['Vulcan'], as yet unobserved, might be the cause. A swarm of intramercurial asteroids was not ruled out, he believed.

Leverrier's figure of $38''$ was soon corrected to its present value of $43''$, but the difficulty for Newtonian explanations of planetary astronomy was still in place 65 years later, when Einstein finally managed to provide a general relativistic explanation "without special assumptions" (such as Vulcan)—an explanation which was soon accepted as strong confirmation for the GTR.

In the diagram above, the left-hand path, $ur \mapsto prn \mapsto new$, represents an expansion of the account in sec. **2.2** of Huygens's "principal" case ("*prn*" for "principal"), in which the confirming phenomenon is verified *after* having been predicted via the hypothesis which its truth confirms. The "*ur*" probability distribution, indicated schematically at the top, represents a time before C was known to follow from H. To accomodate that discovery the "*b*" in the *ur*-distribution is moved left and added to the "*a*" to obtain the top row of the *prn* distribution. The reasoning has two steps. First: Since we now know that H is true, C must be true as well. Therefore $prn(H \wedge \neg C)$ is set equal to 0. Second: Since implying a prediction that may well be false neither confirms nor infirms H, the *prn* probability of H is to remain at its *ur*-value even though the probability of $H \wedge \neg C$ has been nullified. Therefore the probability of $H \wedge C$ is increased to $prn(H \wedge C) = a + b$. And this *prn* distribution is where the Huygens on light example in sec. **2.2** *begins*, leading us to the bottom distribution, which assigns the following odds to H:

$$\frac{new(H)}{new(\neg H)} = \frac{a+b}{c}.$$

In terms of the present example this *new* distribution is what Einstein arrived at from the distribution labelled *old*, in which

$$\frac{old(H)}{old(\neg H)} = \frac{a}{c}.$$

The rationale is

> **Commutativity:** The *new* distribution is the same, no matter if the observation or the explanation comes first.

Now the Bayes factor gives the clearest view of the transition $old(H) \mapsto new(H)$:[20]

$$\beta(H : \neg H) = \frac{a+b}{c} / \frac{a}{c} = 1 + \frac{ur(\neg C|H)}{ur(C|H)}.$$

Thus the new explanation of old news C increases the odds on H by a factor of (1+ the *ur*-odds against C, given H). Arguably, this is very much greater than 1 since, in a notation in which $C = C_{43}$ = an advance of $(43 \pm .5)''/c$ and similarly for other C_i's, $\neg C$ is a disjunction of very many "almost" incompatible disjuncts: $\neg C = \ldots C_{38} \vee C_{39} \vee C_{40} \vee C_{41} \vee C_{42} \vee C_{44} \ldots$[21] That's the good news.[22]

And on the plausible assumption[23] of *ur*-independence of H from C, the formula for β becomes even simpler:

$$\beta(H : \neg H) = 1 + \frac{ur(\bar{C})}{ur(C)}$$

The bad news is that *ur* and *prn* are new constructs, projected into a mythical methodological past. But maybe that's not so bad. As we have just seen in a very sketchy way, there seem to be strong reasons for taking the ratio $ur(\neg C|H)/ur(C|H)$ to be very large. This is clearer in the case of *ur*-independence of H and C:

Example 2, **Ur-independence.** If Wagner's independence assumption is generally persuasive, the physics community's *ur*-odds against C ("C_{43}") will be very high since, behind the veil of *ur*-ignorance, there is a large array of C_i's which, to a first approximation, are equiprobable. Perhaps, 999 of them? Then $\beta(H : \neg H) \geq 1000$, and the weight of evidence (**2.1**) is $w(H : \neg H) \geq 3$.

[20] The approach here is taken from Carl Wagner, "Old evidence and new explanation", *PSA 2000* (J. A. Barrett and J. McK. Alexander, eds.), part 1, pp. S165–S175 (Proceedings of the 2000 biennial meeting of the Philosophy of Science Association, supplement to *Philosophy of Science* **68** [2001]), which reviews and extends earlier work, "Old evidence and new explanation I" and "Old evidence and new explanation II" in *Philosophy of Science* **64**, No. 3 (1997) 677–691 and **66** (1999) 283–288.

[21] "Almost": $ur(C_i \wedge C_j) = 0$ for any distinct integers i and j.

[22] "More [*ur*]danger, more honor." See Pólya, quoted in sec. **2.2** here.

[23] Carl Wagner's, again.

2.6 Supplements

1. "Someone is trying decide whether or not T is true. He notes that T is a consequence of H. Later he succeeds in proving that H is false. How does this refutation affect the probability of T?"[24]
2. "We are trying to decide whether or not T is true. We derive a sequence of consequences from T, say $C_1, C_2, C_3, \ldots$. We succeed in verifying C_1, then C_2, then C_3, and so on. What will be the effect of these successive verifications on the probablity of T?"
3. *Fallacies of "Yes/No" Confirmation.*[25] Each of the following plausible rules is unreliable. Find counterexamples—preferably, simple ones.
   (a) If D confirms T, and T implies H, then D confirms H.
   (b) If D confirms H and T separately, it must confirm their conjunction, TH.
   (c) If D and E each confirm H, then their conjunction, DE, must also confirm H.
   (d) If D confirms a conjunction, TH, then it cannot infirm each conjunct separately.
4. *Hempel's Ravens.* It seems evident that black ravens confirm (H) "All ravens are black" and that nonblack nonravens do not. Yet H is logically equivalent to "All nonravens are nonblack". Use probabilistic considerations to resolve this paradox of "yes/no" confirmation.[26]
5. *Wagner III* (sec **2.5**).[27] Carl Wagner extends his treatment of old evidence newly explained to cases in which updating is by generalized conditioning on a countable sequence $\mathcal{C} = \langle C_1, C_2, \ldots \rangle$ where none of the updated probabilies need be 1, and to cases in

[24] This problem and the next are from George Polya, "Heuristic reasoning and the theory of probability", *American Mathematical Monthly* **48** (1941) 450–465.

[25] These stem from Carl G. Hempel's "Studies in the logic of confirmation", *Mind* **54** (1945) 1–26 and 97–121, which is reprinted in his *Aspects of Scientific Explanation*, New York, The Free Press, 1965.

[26] The paradox was first floated (1937) by Carl G. Hempel, in an abstract form: See pp. 50–51 of his *Selected Philosophical Essays*, Cambridge University Press, 2000. For the first probabilistic solution, see "On confirmation" by Janina Hosiasson-Lindenbaum, *The Journal of Symbolic Logic* **5** (1940) 133–148. For a critical survey of more recent treatments, see pp. 69–73 of John Earman's *Bayes or Bust?*

[27] This is an easily detachable bit of Wagner's "Old evidence and new explanation III" (Wagner, 2001, cited above in sec. **2.5**).

which each of the $\mathcal{C}$'s "follows only probabilistically" from H in the sense that the conditional probabilities of the $\mathcal{C}$'s, given H, are high but < 1. Here we focus on the simplest case, in which $\mathcal{C}$ has two members, $\mathcal{C} = \langle C, \neg C \rangle$. Where (as in the diagram in sec. **2.6**) updates are not always from *old* to *new*, we indicate the prior and posterior probability measures by a subscript and superscript: β^{post}_{prior}. In particular, we shall write $\beta^{new}_{old}(H : \neg H)$ explicitly, instead of $\beta(H : \neg H)$ as in (3) and (4) of sec. **2.6.** And we shall use "β^*" as shorthand: $\beta^* = \beta^{old}_{ur}(C : \neg C)$.

(a) Show that in the framework of sec. **2.6** COMMUTATIVITY is equivalent to UNIFORMITY:

$$\beta^{new}_{old}(A : A') = \beta^{prn}_{ur}(A : A') \text{ if } A, A' \in \{H \wedge C, H \wedge \neg C, \neg H \wedge C, \neg H \wedge \neg C\}.$$

(b) Show that uniformity holds whenever the updates $ur \mapsto old$ and $prn \mapsto new$ are both by generalized conditioning. Now verify that where uniformity holds, so do the following two formulas:

(c) $$\beta^{new}_{ur}(H : \neg H) = \frac{(\beta^* - 1)prn(C|H) + 1}{(\beta^* - 1)prn(C|\neg H) + 1}$$

(d) $$\beta^{old}_{ur}(H : \neg H) = \frac{(\beta^* - 1)ur(C|H) + 1}{(\beta^* - 1)ur(C|\neg H) + 1}$$

And show that, given uniformity,

(e) if H and C are *ur*-independent, then $old(H) = ur(H)$ and, therefore,

(f) $$\beta^{new}_{old}(H : \neg H) = \frac{(\beta^* -)prn(C|H) + 1}{(\beta^* - 1)prn(C|\neg H) + 1}, \text{ so that}$$

(g) Given uniformity, if H and C are *ur*-independent then
   (1) $\beta^{new}_{old}(H : \neg H)$ depends only on β^*, $prn(C|H)$ and $prn(C|\neg H)$;
   (2) if $\beta > 1$ and $prn(C|H) > prn(C|\neg H)$, then $\beta^{new}_{old}(H : \neg H) > 1$;
   (3) $\beta^{new}_{old}(H : \neg H) \rightarrow \frac{prn(C|H)}{prn(C|\neg H)}$ as $\beta \rightarrow \infty$.

6. *Shimony on Holt-Clauser.*[28] "Suppose that the true theory is local hidden variables, but Clauser's apparatus [which supported

[28] Personal communication, 12 Sept. 2002.

QM] was faulty. Then you have to accept that the combination of systematic and random errors yielded (within the rather narrow error bars) the quantum mechanical prediction, which is a definite number. The probability of such a coincidence is very small, since the part of experimental space that agrees with a numerical prediction is small. By contrast, if quantum mechanics is correct and Holt's apparatus is faulty, it is not improbable that results in agreement with Bell's inequality (hence with local hidden variables theories) would be obtained, because agreement occurs when the relevant number falls within a rather large interval. Also, the errors would usually have the effect of disguising correlations, and the quantum mechanical prediction is a strict correlation. Hence, good Bayesian reasons can be given for voting for Clauser over Holt, even if one disregards later experiments devised in order to break the tie."

Compare: Relative ease of the Rockies eventually wearing down to Adirondacks size, as against improbability of the Adirondacks eventually reaching the size of the Rockies.

3

Probability Dynamics; Collaboration

We now look more deeply into the matter of learning from experience, where a pair of probability assignments represents your judgments before and after you change your mind in response to the result of experiment or observation. We start with the simplest, most familiar mode of updating, which will be generalized in sec. **3.2** and applied in sec. **3.3** and **3.4** to the problem of learning from other people's experience-driven probabilistic updating.

3.1 Conditioning

Suppose your judgments were once characterized by an assignment ("*old*") which describes all of your conditional and unconditional probabilities as they were then. And suppose you became convinced of the truth of some data statement D. That would change your probabilistic judgments; they would no longer corrrespond to your prior assignment *old*, but to some posterior assignment *new*, where you are certain of D's truth.

Certainty: $new(D) = 1$

How should your *new* probabilities for hypotheses H other than D be related to your *old* ones?

The simplest answer goes by the name of "conditioning" (or "conditionalization") on D.

Conditioning: $new(H) = old(H|D)$

This means that your new unconditional probability for any hypothesis H will simply be your old conditional probability for H given D.

When is it appropriate to update by conditioning on D? It is easy to see—once you think of it—that certainty about D is not enough.

EXAMPLE 1, **The red queen.** You learn that a playing card, drawn from a well-shuffled, standard 52-card deck, is a heart: your $new(\heartsuit) = 1$. Since you know that all hearts are red, this means that your $new(\text{red}) = 1$ as well. But since $old(\text{Queen} \wedge \heartsuit | \heartsuit) = \frac{1}{13}$ whereas $old(\text{Queen} \wedge \heartsuit | \text{red}) = \frac{1}{26}$, you will have different new probabilities for the card's being the Queen of hearts, depending on which certainty you update on, $\heartsuit$ or *red.*

Of course, in example 1 we are in no doubt about which certainty you should condition on: it is the more informative one, $\heartsuit$. And if what you actually learned was that the card was a court card as well as a heart, then that is what you ought to update on, so that your $new(Q \wedge \heartsuit) = old(Q \wedge \heartsuit | (\text{Ace} \vee \text{King} \vee \text{Queen}) \wedge \heartsuit)) = \frac{1}{3}$. But what if—although the thing you learned with certainty was indeed that the card was a heart—you also saw something you can't put into words, that gave you reason to suspect, uncertainly, that it was a court card?

EXAMPLE 2, **A brief glimpse** determines your new probabilities as 1/20 for each of the ten numbered hearts, and as 1/6 for each of the J, K and Q. Here, where some of what you have learned is just probabilistic, conditioning on your most informative new certainty does not take all of your information into account. In sec. **3.2** we shall see how you might take such uncertain information into account; but meanwhile, it is clear that ordinary conditioning is not always the way to go.

What is the general condition under which it makes sense for you to plan to update by conditioning on D if you should become sure that D is true? The answer is confidence that your conditional probabilities given D will not be changed by the experience that makes you certain of D's truth.

Invariant Conditional Probabilities:

$$\text{(1)} \quad \text{For all } H, new(H|D) = old(H|D)$$

Since $new(H|D) = new(H)$ when $new(D) = 1$, it is clear that certainty and invariance together imply conditioning. And conversely, conditioning implies both certainty and invariance.[1]

An equivalent invariance is that of your unconditional odds between different ways in which D might be true.

[1] To obtain Certainty, set $H = D$ in Conditioning. Now invariance follows from Conditioning and Certainty since $new\ H = new(H|D)$ when $new\ D = 1$.

Invariant Odds:

$$\text{(2)} \quad \text{If } A \text{ and } B \text{ each imply } D, \quad \frac{new(A)}{new(B)} = \frac{old(A)}{old(B)}.$$

And another is invariance of the "probability factors" $\pi(A)$ by which your old probabilities of ways of D's being true can be multiplied in order to get their new probabilities.

Invariant Probability Factors:

$$\text{(3)} \quad \text{If } A \text{ implies } D \text{ then } \pi(A) = \pi(D)$$

$$\text{—where } \pi(A) =_{df} \frac{new(A)}{old(A)}.$$

Exercise. Show that (1) implies (3), which implies (2), which implies (1).

There are special circumstances in which the invariance condition dependably holds:

Example 3, **The statistician's stooge.**[2] Invariance will surely hold in the variant of example 1 in which you do not see the card yourself, but have arranged for a trusted assistant to look at it and tell you simply whether or not it is a heart, with no further information. Under such conditions your unconditional probability that the card is a heart changes to 1 in a way that cannot change any of your conditional probabilities, given ♡.

3.2 Generalized Conditioning[3]

Certainty is quite demanding. It rules out not only the far-fetched uncertainties associated with philosophical skepticism, but also the familiar uncertainties that affect real empirical inquiry in science and everyday life. But it is a condition that turns out to be dispensable: As long as invariance holds, updating is valid by a generalization of conditioning to which we now turn. We begin with the simplest case, of a yes/no question.

If invariance holds relative to each answer $(D, \neg D)$ to a yes/no question, and something makes you (say) 85% sure that the answer is 'yes',

[2] I. J. Good's term.

[3] Also known as *probability kinematics* and *Jeffrey conditioning.* For a little more about this, see Persi Diaconis and Sandy Zabell, 'Some alternatives to Bayes's rule' in *Information Pooling and Group Decision Making,* Bernard Grofman and Guillermo Owen (eds.), Greenwich, Conn., and London, England, JAI Press, pp. 25–38. For much more, see 'Updating subjective probability' by the same authors, *Journal of the American Statistical Association* **77** (1982) 822–830.

a way of updating your probabilities is still available for you. Invariance with respect to both answers means that for each hypothesis, H, $new(H|D) = old(H|D)$ and $new(H|\neg D) = old(H|\neg D)$. The required updating rule is easily obtained from the law of total probability in the form

$$new(H) = new(H|D)new(D) + new(H|\neg D)new(\neg D).$$

Rewriting the two conditional probabilities in this equation via invariance relative to D and to $\neg D$, we have the following updating scheme:

$$new(H) = old(H|D)new(D) + old(H|\neg D)new(\neg D).$$

More generally, with any countable partition of answers, the applicable rule of total probability has one term on the right for each member of the partition. If the invariance condition holds for each answer, D_i, we have the updating scheme $new(H) = old(H|D_1)\ new(D_1) + old(H|D_2)\ new(D2) + \ldots,$

"Probability Kinematics":

$$new(H) = \sum_{i=1}^{n} old(H|D_i)new(D_i).$$

This is equivalent to invariance with respect to every answer: $new(H|D_i) = old(H|D_i)$ for $i = 1, \ldots, n$.

EXAMPLE 4, **A consultant's prognosis.** In hopes of settling on one of the following diagnoses, Dr. Jane Doe, a histopathologist, conducts a microscopic examination of a section of tissue surgically removed from a tumor. She is sure that exactly one of the three is correct:

D_1 = Islet cell carcinoma, D_2 = Ductal cell carcinoma, D_3 = Benign tumor.

Here $n = 3$, so (using accents to distinguish her probability assignments from yours) her $new'(H)$ for the prognosis H of 5-year survival will be

$$old'(H|D_1)new'(D_1) + old'(H|D_2)new'(D_2) + old'(H|D_3)new'(D_3).$$

Suppose that, in the event, the examination does not drive her probability for any of the diagnoses to 1, but leads her to assign $new'(D_i) = \frac{1}{3}, \frac{1}{6}, \frac{1}{2}$ for $i = 1, 2, 3$. Suppose, further, that her conditional probabilities for H given the diagnoses are unaffected by her examination: $old'(H|D_i) = new'(H|D_i) = \frac{4}{10}, \frac{6}{10}, \frac{9}{10}$. Then by probability kinematics her new probability for 5-year survival will be a weighted average

$new'(H) = \frac{41}{60}$ of the values that her $old'(H)$ would have had if she had been sure of the three diagnoses, where the weights are her new probabilities for those diagnoses.

Note that the calculation of $new'(H)$ makes no use of Dr. Doe's old probabilities for the diagnoses. Indeed, her prior old' may have been a partially defined function, assigning no numerical values at all to the D_i's. Certainly her $new'(D_i)$'s will have arisen through an interaction of features of her prior mental state with her new experiences at the microscope. But the results $new'(D_i)$ of that interaction are *data* for the kinematical formula for updating $old'(H)$; the formula itself does not compute those results.

3.3 Probabilistic Observation Reports

We now move outside the native ground of probability kinematics into a region where your new probabilities for the D_i are to be influenced by someone else's probabilistic observation report.[4] You are unlikely to simply adopt such an observer's updated probabilities as your own, for they are necessarily a confusion of what the other person has gathered from the observation itself, which you would like to adopt as your own, with that person's prior judgmental state, for which you may prefer to substitute your own.

We continue in the medical setting. Suppose you are a clinical oncologist who wants to make the best use you can of the observations of a histopathologist whom you have consulted. Notation: *old* and *new* are your probability assignments *before* and *after* the histopathologist's observation has led her to change her probability assignment from old' to new'.

If you do simply adopt the expert's new probabilities for the diagnoses, setting your $new(D_i) = new(D_i)$ for each i, you can update by probability kinematics even if you had no prior diagnostic opinions $old(D_i)$ of your own; all you need are her new $new'(D_i)$'s and your invariant conditional prognoses $old(H|D_i)$. But suppose you have your own priors, $old(D_i)$, which you take to be well-founded, and although you have high regard for the pathologist's ability to interpret histographic slides, you view her prior probabilities $old'(D_i)$ for the various diagnoses as arbitrary and uninformed: She has no background information about the patient, but for the purpose of formulating her report she has adopted

[4] See Wagner (2001) and Wagner, 'Probability Kinematics and Commutativity', *Philosophy of Science* 69 (2002) 266–278.

certain numerically convenient priors to update on the basis of her observations. It is not from the $new'(D_i)$ themselves but from the updates $old'(D_i) \mapsto new'(D_i)$ that you must extract the information contributed by her observation. How?

Here is a way: Express your $new(D_i)$ as a product $\pi'(D_i)old(D_i)$ in which the factor $\pi'(D_i)$ is constructed by combining the histopathologist's probability factors for the diagnoses with your priors for them, as follows.[5]

$$(1) \qquad \pi(D_i) := \frac{\pi'(D_i)}{\sum_i \pi'(D_i)old(D_i)} \qquad \text{(Your } D_i \text{ probability factor)}$$

Now, writing your $new(D_i) = \pi'(D_i)old'(D_i)$ in the formula in sec. **2.2** for updating by probability kinematics, and applying the product rule, we have

$$(2) \qquad new(H) = \sum_i \pi'(D_i)old(H \wedge D_i)$$

as your updated probability for a prognosis, in the light of the observer's report and your prior probabilities, and

$$(3) \qquad \frac{new(H_1)}{new(H_2)} = \frac{\sum_i \pi'(D_i)old(H_1 \wedge D_i)}{\sum_i \pi'(D_i)old(H_2 \wedge D_i)}$$

as your updated odds between two prognoses.[6]

EXAMPLE 5, **Using the consultant's diagnostic updates.** In example 4 the histopathologist's unspecified priors were updated to new values $new'(D_i) = \frac{1}{3}, \frac{1}{6}, \frac{1}{2}$ for $i = 1, 2, 3$ by her observation. Now if all her $old'(D_i)$'s were $\frac{1}{3}$, her probability factors would be $\pi'(D_i) = 1, \frac{1}{2}, \frac{3}{2}$; if your $old(D_i)$'s were $\frac{1}{4}, \frac{1}{2}, \frac{1}{4}$, your diagnostic probability factors (1) would be $\pi(D_i) = \frac{4}{3}, \frac{2}{3}, 2$; and if your $old(H \wedge D_i)$'s were $\frac{3}{16}, \frac{3}{32}, \frac{3}{32}$ then by (2) your $new(H)$ would be $\frac{1}{2}$ as against $old(H) = \frac{3}{8}$, so that your π(5-year survival) $= \frac{4}{3}$.

It is noteworthy that formulas (1)–(3) remain valid when the probability factors are replaced by anchored odds (or "Bayes") factors:

[5] Without the normalizing denominator in (1), the sum of your $new(D_i)$'s might turn out to be $\neq 1$.

[6] Combine (1) and (2), then cancel the normalizing denominators to get (3).

$$(4) \qquad \beta(D_i : D_1) = \frac{new D_i}{new(D_1)} / \frac{old(D_i)}{old(D_1)} = \frac{\pi(D_i)}{\pi(D_1)} \qquad (D_i : D_1 \text{ Bayes factor})$$

This is your Bayes factor for D_i against D_1; it is the number by which your old odds on D_1 against D_1 can be multiplied in order to get your new odds; it is what remains of your new odds when the old odds have been factored out. Choice of D_1 as the anchor diagnosis with which all the D_i are compared is arbitrary. Since odds are arbitrary up to a constant multiplier, any fixed diagnosis D_k would do as well as D_1, for the Bayes factors $\beta(D_i : D_k)$ are all of form $\beta(D_i : D_1) \times c$, where the constant c is $\beta(D_1 : D_k)$.

Bayes factors have wide credibility as probabilistic observation reports, with prior probabilities "factored out".[7] But probability factors carry some information about priors, as when a probability factor of 2 tells us that the old probability must have been $\leq 1/2$ since the new probability must be ≤ 1. This residual information about priors is neutralized in the context of formulas (1)–(3) above, which remain valid with anchored Bayes factors in place of probability factors. This matters (a little) because probability factors are a little easier to compute than Bayes factors, starting from old and new probabilities.

3.4 Updating Twice: Commutativity

Here we consider the outcome of successive updating on the reports of two different experts—say, a histopathologist ($'$) and a radiologist ($''$)—assuming invariance of your conditional probabilities relative to both experts' partitions. Question: If you update twice, should order be irrelevant?

$$\text{Should } \overset{\prime}{\mapsto}\overset{\prime\prime}{\mapsto} = \overset{\prime\prime}{\mapsto}\overset{\prime}{\mapsto}?$$

The answer depends on particulars of

(1) the partitions on which $\overset{\prime}{\mapsto}$ and $\overset{\prime\prime}{\mapsto}$ are defined;
(2) the mode of updating (by probabilities? Bayes factors?); and
(3) your starting point, P.

[7] In a medical context, this goes back at least as far as Schwartz, Wolfe, and Pauker, "Pathology and Probabilities: a new approach to interpreting and reporting biopsies," *New England Journal of Medicine* **305** (1981) 917–923.

3.4.1 Updating on Alien Probabilities for Diagnoses

Apropos of (2), suppose you accept two new probability assignments to one and the same partition, in turn.

- Can order matter?

  Certainly. Since in both updates your $new(H)$ depends only on your invariant conditional probabilities $old(H|D_i)$ and the $new(D_i)$'s that you adopt from that update's expert, the second assignment simply replaces the first.
- When is order immaterial?

  When there are two partitions, and updating on the second leaves probabilities of all elements of the first unchanged—as happens, e.g., when the two partitions are independent relative to your *old.*[8]

3.4.2 Updating on Alien Factors for Diagnoses[9]

Notation: Throughout each formula, $f = \beta$ or $f = \pi$, as you wish.

In updating by Bayes or probability factors f for diagnoses as in **3.3**, order cannot matter.[10]

EXAMPLE 6, **One partition.** Adopting as your own both a pathologist's factors $f_i^{'}$ and a radiologist's factors $f_i^{''}$ on the same partition—in either order—you come to the same result: Your overall factors will be products $f_i^{'} f_i^{''}$ of the pathologist's and radiologist's factors. Your final probabilities for the diagnoses and for the prognosis B will be

$$new(D_i) = old(D_i)f,\ Q(H) = \sum_i old(H \wedge D_i)f$$

where f is your normalized factor, constructed from the alien $f^{'}$ and$^{''}$:

$$f = \frac{f_i^{'} f_i^{''}}{\sum_i old(A_i) f_i^{'} f_i^{''}}$$

EXAMPLE 7, **Two partitions.** A pathologist's, $\overset{'}{\mapsto}$, with partition

[8] For more about this, see Diaconis and Zabell (1982), esp. 825–826.
[9] For more about this, see Carl Wagner's 'Probability Kinematics and Commutativity', *Philosophy of Science* **69** (2002) 266–278.
[10] Proofs are straightforward. See pp. 52–64 of Richard Jeffrey, Petrus Hispanus Lectures 2000: *After Logical Empiricism*, Sociedade Portuguesa de Filosofia, Edições Colibri, Lisbon (2002) (edited, translated, and introduced by António Zilhão. Also in Portugese: *Depois do empirismo lógico*.)

$\{D_i'\}$ and factors f_i' $(i = 1, \ldots, m)$, and a radiologist's, $\overset{''}{\mapsto}$, with partition $\{D_j''\}$ and factors f_j'' $(j = 1, \ldots, n)$. These must commute, for in either order they are equivalent to a single mapping, $\mapsto$, with partition $\{D_i' \wedge D_j'' \,|\, old(A_i' \wedge A_j'') > 0\}$ and factors $f_i' f_j''$. Now in terms of your normalization

$$f_{i,j} = \frac{f_i' f_j''}{\sum_{i,j} old(D_i' \wedge D_j'') f_i' f_j''}$$

of the alien factors, your updating equations on the partition elements $D_i' \wedge D_j''$ can be written $new(D_i' \wedge D_j'') = f_{i,j}\, old(D_i' \wedge D_j'')$. Then your new probability for the prognosis will be

$$new(H) = \sum_{i,j} f_{i,j}\, old(H \wedge D_i' \wedge D_j'').$$

3.5 Softcore Empiricism

Here is a sample of mid-20th century hardcore empiricism:

> Subtract, in what we say that we see, or hear, or otherwise learn from direct experience, *all that conceivably could be mistaken;* the remainder is the given content of the experience inducing this belief. If there were no such hard kernel in experience—e.g., what we see when we think we see a deer but there is no deer—then the word 'experience' would have nothing to refer to.[11]

Hardcore empiricism puts some such "hard kernel" to work as the "purely experiential component" of your observations, about which you cannot be mistaken. Lewis himself thinks of this hard kernel as a proposition, since it has a truth value (i.e., true), and has a subjective probability for you (i.e., 1). Early and late, he argues that the relationship between your irrefutable kernel and your other empirical judgments is the relationship between fully believed premises and uncertain conclusions, which have various probabilities conditional upon those premises. Thus, in 1929 (*Mind and the World Order,* pp. 328–329) he holds that

[11] C. I. Lewis, *An Analysis of Knowledge and Valuation*, LaSalle, Illinois, Open Court, 1947, pp. 182–183. Lewis's emphasis.

> the immediate premises are, very likely, themselves only probable, and perhaps in turn based upon premises only probable. Unless this backward-leading chain comes to rest finally in certainty, no probability-judgment can be valid at all. [...] Such ultimate premises ... must be actual given data for the individual who makes the judgment.

And in 1947 (*An Analysis of Knowledge and Valuation*, p. 186):

> If anything is to be probable, then something must be certain. The data which themselves support a genuine probability, must themselves be certainties. We do have such absolute certainties in the sense data initiating belief and in those passages of experience which later may confirm it.

In effect, Lewis subscribes to Carnap's view[12] of inductive probability as prescribing, as your current subjective probability for a hypothesis H, your $old(H|D_1 \wedge \ldots \wedge D_t)$, where the D_i are your fully believed data sentences from the beginning (D_1) to date (D_t) and old is a probability assignment that would be appropriate for you to have prior to all experience.

Lewis's arguments for this view seem to be based on the ideas, which we have been undermining in this chapter, that (a) conditioning on certainties is the only rational way to form your degrees of belief, and (b) if you are rational, the information encoded in your probabilities at any time is simply the conjunction of all your hardcore data sentences up to that time. But perhaps there are softcore, probabilistic data sentences on which simple conditioning is possible to the same effect as probability kinematics:

EXAMPLE 8, **Is the shirt blue or green?** There are conditions (blue/green color-blindness, poor lighting) under which what you see can lead you to adopt probabilities—say, $\frac{2}{3}$, $\frac{1}{3}$—for blue and green, where there is no hardcore experiential proposition E you can cite for which your $old(Blue|E) = \frac{2}{3}$ and $old(Green|E) = \frac{1}{3}$. The required softcore experiential proposition $\mathcal{E}$ would be less accessible than Lewis's 'what we *see* when we think we see a deer but there is no deer', since what we think is not that the shirt is blue, but that $new(\text{blue}) = \frac{2}{3}$ and $new(\text{green}) = \frac{1}{3}$. With $\mathcal{E}$ as the proposition $\boxed{new(\text{blue}) = \frac{2}{3} \wedge new(\text{green}) = \frac{1}{3}}$, it is possible to expand the domain of the function *old so* as to allow conditioning

[12] See, e.g., Rudolf Carnap, *The Logical Foundations of Probability*, University of Chicago Press, 1950, 1964.

on $\mathcal{E}$ in a way that yields the same result you would get via probability kinematics. Formally, this $\mathcal{E}$ behaves like the elusive hardcore E at the beginning of this example.

This is how it works.[13] For an n-fold partition, where

$$\mathcal{E} = \boxed{new(D_1) = d_1 \wedge \ldots \wedge new(D_n) = d_n}.^{14}$$

If

(1) $old(D_i | \boxed{new(D_i) = d_i}) = d_i$

and

(2) $old(H | D_i \wedge \mathcal{E}d) = old(H | D_i)$

then

(3) $old(H | \mathcal{E}) = d_1\, old(H | D_1) + \ldots + d_n\, old(H | D_n)$.

In the presence of the certainty condition, $new(\mathcal{E}) = \infty$, the invariance condition $new(H | \mathcal{E}) = old(H | \mathcal{E})$ reduces to $new(H) = old(H | \mathcal{E})$ and we have probability kinematics,

(4) $new(H) = d_1\, old(H | D_1) + \ldots d_n\, old(H | D_n)$.

Perhaps $\mathcal{E}$ could be regarded as an experiential proposition in some soft sense. It may be thought to satisfy the certainty and invariance conditions, $new(\mathcal{E}) = 1$ and $new(H | \mathcal{E}) = old(H | \mathcal{E})$, and it does stand outside the normal run of propositions to which we assign probabilities, as do the presumed experiential propositions with which we might hope to cash those heavily context-dependent epistemological checks that begin with 'looks'. The price of that trip would be a tricky expansion of the probability assignment, from the ("first-order") objective propositions such as diagnoses and prognoses that figure in the kinematical update formula to subjective propositions of the second order ($\mathcal{E}$ and $\boxed{new(D_i) = d_i}$) and third ($\boxed{old(\mathcal{E}) = 1}$) and beyond.[15]

[13] Brian Skyrms, *Causal Necessity*, New Haven, Conn. University Press, Yale 1980, Appendix 2.

[14] Such boxes are equivalent to fore-and-aft parentheses, and easier to parse.

[15] For a model theory of this process, with references to earlier work of Haim Gaifman and others, see Dov Samet, "Bayesianism without Learning," *Research in Economics* **53** (1999) 227–242.

4
Expectation Primer

Your **expectation**, $ex(X)$, of an unknown number, X, is usually defined as a weighted average of the numbers you think X might be, in which the weights are your probabilities for X's being those numbers: ex is usually defined in terms of pr. But in many ways the opposite order is preferable, with pr Defined in terms of ex as in this chapter.[1]

4.1 Probability and Expectation

If you see only finitely many possible answers, $x_1, \ldots, x_n$, to the question of what number X is, your expectation can be defined as follows:

$$(1) \qquad ex(X) = \sum_{i=1}^{n} p_i x_i \qquad \text{—where } p_i = pr(X = x_i).$$

EXAMPLE, **X = The year Turing died.** If I am sure it was one of the years from 1951 to 1959, with equal probabilities, then my $ex(X)$ will be $\frac{1951}{9} + \frac{1952}{9} + \ldots + \frac{1959}{9}$, which works out to be 1954.5. And if you are sure Turing died in 1954, then for you $n = 1$, $p_1 = 1$ and $ex(X) =$ 1954.

These numbers X are called **random variables, R.V.'s.** Since an R.V. is always some particular number, random variables are really constants—known or unknown. The phrase 'the year Turing died' is

[1] —and as in the earliest texbook of probability theory: Christiaan Huygens, *De Ratiociniis in Ludo Aleae*, 1657 ("On Calculating in Games of Luck"). Notable 20th-century expectation-based treatments are Bruno de Finetti's *Teoria delle Probabilità*, v. 1, Einaudi, 1970 (= *Theory of Probability,* Wiley, 1975) and Peter Whittle's *Probability via Expectation,* Springer, 4th ed., 2000.

a constant which, as a matter of empirical fact, denotes the number 1954, quite apart from whether you or I know it.

We could have taken expectation as the fundamental concept, and defined probabilities as expectations of particularly simple R.V.'s, called 'indicators'. The ***indicator*** of a hypothesis H is a constant, $\boldsymbol{I_H}$, which is 1 if H is true and 0 if H is false. Now probabilities are expectations of indicators,

(2) $pr(H) = ex(I_H)$.

This is equivalent to definition (1).

Definition (2) is quite general. In contrast, definition (1) works only in cases where you are sure that X belongs to a particular finite list of numbers. Of course, definition (1) can be generalized, but this requires relatively advanced mathematical concepts which definition (2) avoids, or, anyway, sweeps under the carpet.[2]

Observe that expectations of R.V.'s need not be values those R.V.'s can have. In the Turing example my expectation of the year of Turing's death was 1954.5, which is not the number of a year. Nor need your expectation of the indicator of past life on Mars be one of the values, 0 or 1, that indicators can assume; it may well be $\frac{1}{10}$, as in the story in sec. **1.1.**

4.2 Conditional Expectation

Just as we defined your conditional probabilities as your buying-or-selling prices for tickets that represent conditional bets, so we define your conditional expectation of a random variable X given truth of a statement H as your buying-or-selling price for the following ticket:

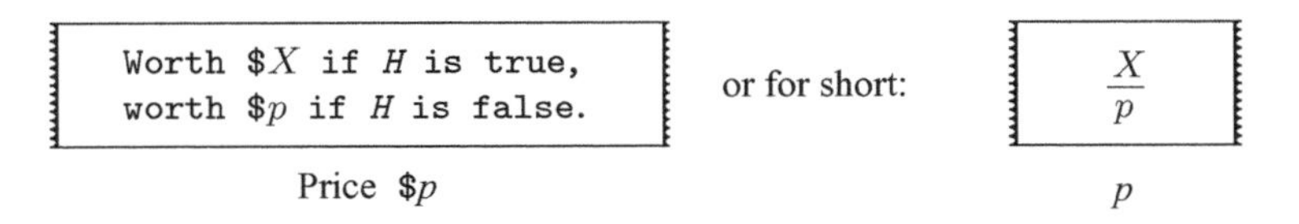

We write '$ex(X|H)$' for your conditional expectation of X given H.

The following rule might be viewed as defining conditional expectations as quotients of unconditional ones when $pr(H) \neq 0$, just as the

[2] For example, the Riemann-Stieltjes integral: If you think X might be any real number from a to b, then (1) becomes: $ex(X) = \int_a^b x\, dF(x)$, where $F(x) = pr(X < x)$.

quotient rule for probabilities might be viewed as defining $pr(G|H)$ as $pr(GH)/pr(H)$.

$$\textbf{Quotient Rule:} \quad ex(X|H) = \frac{ex(X \cdot I_H)}{ex(I_H)} = \frac{ex(X \cdot I_H)}{pr(H)} \text{ if } pr(H) \neq 0.$$

Of course this can be written as follows, without the condition $pr(H) \neq 0$.

$$\textbf{Product Rule:} \quad ex(X \cdot I_H) = ex(X|H) \cdot pr(H)$$

A "Dutch book" consistency argument can be given for this product rule rather like the one given in sec. **1.4** for the probability product rule: Consider the following five tickets.

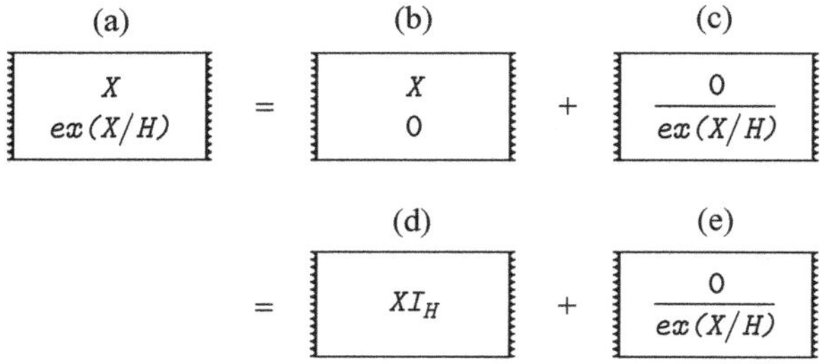

(a) $\left[\frac{X}{ex(X/H)}\right]$ = (b) $\left[\frac{X}{0}\right]$ + (c) $\left[\frac{0}{ex(X/H)}\right]$

= (d) $\left[XI_H\right]$ + (e) $\left[\frac{0}{ex(X/H)}\right]$

Clearly, ticket (a) has the same dollar value as (b) and (c) together on each hypothesis as to whether H is true or false. And (d) and (e) have the same values as (b) and (c), since $X \cdot I_H = \{ {X \text{ if } H \text{ is true} \atop 0 \text{ if } \neg H \text{ is true}}$ and (e) = (c). Then unless your price for (a) is the sum of your prices for (d) and (e), so that the condition $ex(X|H) = ex(X \cdot I_H) + ex(X|H)pr(\neg H)$ is met, you are inconsistently placing different values on the same prospect depending on whether it is described in one or the other of two provably equivalent ways. Now setting $pr(\neg H) = 1 - pr(H)$ and simplifying, the condition boils down to the product rule.

Historical Note. In the 18th century Thomas Bayes defined probability in terms of expectation as follows:

> "The probability of any event is the ratio between the value at which an expectation depending on the happening of the event ought to be computed, and the value of the thing expected upon its happening."[3]

[3] 'Essay toward solving a problem in the doctrine of chances,' *Philosophical Transactions of the Royal Society* **50** (1763) 376, reprinted in *Facsimiles of Two Papers by Bayes*, New York, Hafner, 1963.

This is simply the product rule, solved for $prH = \dfrac{ex(X \cdot I_H)}{ex(X|H)}$. Explanation: $ex(X \cdot I_H)$ is "an expectation [\$ X] depending on the happening of the event" and $ex(X|H)$ is "the value of the thing expected upon its [H's] happening".

4.3 Laws of Expectation

The axioms for expectation can be taken to be the product rule and

(3) **Linearity:** $ex(aX + bY + c) = a\,ex(X) + b\,ex(Y) + c$,

where the small letters stand for *numerals* (= random variables whose values everybody who understands the language is supposed to know).

Three notable special cases of linearity are obtained if we replace 'a, b, c' by '1, 1, 0' (additivity) or '$a, 0, 0$' (proportionality) or '0, 0, c' (constancy, then proportionality with 'c' in place of 'a'):

	Additivity:	$ex(X + Y) = ex(X) + ex(Y)$
(4)	**Proportionality:**	$ex(aX) = a\ ex(X)$
	Constancy:	$ex(\top) = 1$

If n hypotheses H_i are well defined, they have actual truth values (truth or falsehood) even if we do not know what they are. Therefore we can define the **number of truths** among them as $I_{H_1} + \ldots + I_{H_n}$. If you have definite expectations $ex(I_{H_i})$ for them, you will have a definite expectation for that sum, which, by additivity, will be $ex(I_{H_1}) + \ldots + ex(I_{H_1})$.

EXAMPLE 1, **Calibration.** Suppose you attribute the same probability, p, to success on each trial of a certain experiment: $pr(H_1) = \ldots = pr(H_n) = p$. Consider the indicators of the hypotheses, H_i, that the different trials succeed. The number of successes in the n trials will be the unknown sum of the indicators, so the unknown **success rate** will be $\boldsymbol{S_n} = \dfrac{1}{n}\displaystyle\sum_{i=1}^{n} H_i$. Now, by additivity and constancy, your expectation of the success rate must be p.[4]

The term 'calibration' comes from the jargon of weather forecasting; forecasters are said to be well calibrated—say, last year, for rain, at the

[4] $ex(\frac{1}{n}\sum_{i=1}^{n} I_{H_i}) = \frac{1}{n}ex(\sum_{i=1}^{n} I_{H_i}) = \frac{1}{n}ex(\sum_{i=1}^{n} pr(H_i)) = \frac{1}{n}(np) = p$.

level $p = .8$—if last year it rained on about 80% of the days for which the forecaster turns out to have said there would be an 80% chance of rain.[5]

Why Expectations are Additive. Suppose x and y are your expectations of R.V.'s X and Y—say, rainfall in inches during the first and second halves of next year—and z is your expectation for next year's total rainfall, $X + Y$. Why should z be $x + y$? The answer is that in every eventuality about rainfall at the two locations, the value of the first two of these tickets together is the same as the value of the third:

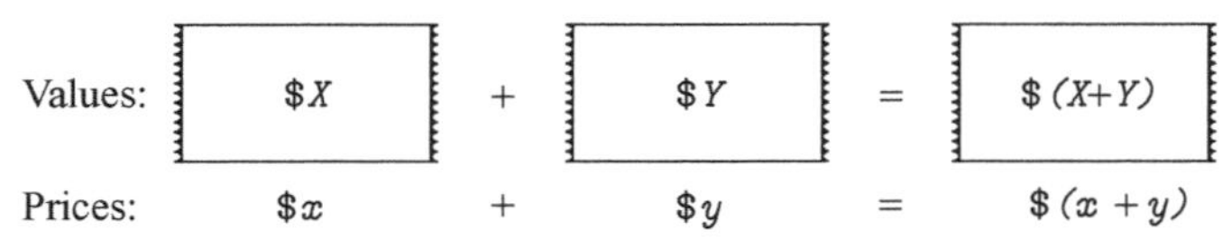

Then unless the prices you would pay for the first two add up to the price you would pay for the third, you are inconsistently placing different values on the same prospect, depending on whether it is described to you in one or the other of two provably equivalent ways.

Typically, a random variable might have any one of a number of values as far as you know. Convexity is a consequence of linearity according to which your expectation for the R.V. cannot be larger than all of those values, or smaller than all of them:

> **Convexity:** If you are sure that X is one of a finite collection of numbers, $ex(X)$ lies in the range from the largest to the smallest to the smallest of them.

Another connection between conditional and unconditional expectations:

> **Law of Total Expectation.** If no two of $H_1, \ldots, H_n$ are compatible and collectively they exhaust $\top$, then
> $$ex(X) = \sum_{i=1}^{n} ex(X|H_i)pr(H_i).$$

Proof. $X = \sum_{i=1}^{n}(X \cdot I_{H_i})$ is an identity between R.V.'s. Apply ex to both sides, then use additivity and the product rule.

When conditions are certainties for you, conditional expectations reduce to unconditional ones:

[5] For more about calibration, etc., see Morris DeGroot and Stephen Feinberg, "Assessing Probability Assessors: Calibration and Refinement," in Shanti S. Gupta and James O. Berger (eds.), *Statistical Decision Theory and Related Topics III*, vol. 1, New York, Academic Press, 1982, pp. 291–314.

Certainty: $ex(X|H) = ex(X)$ if $pr(H) = 1$.

Note that in a context of form $ex(\cdots Y \cdots |Y = X)$ it is always permissible to rewrite Y as X at the left. Example: $ex(Y^2|Y = 2X) = ex(4X^2|Y = 2X)$.

Applying Conditions:

$$ex(\cdots Y \cdots |Y = X) = ex(\cdots X \cdots |Y = X) \qquad \textbf{(OK!)}$$

But we cannot generally discharge a condition $Y = X$ by rewriting Y as X at the left *and dropping the condition.* Example: $ex(Y^2|Y = 2X)$ cannot be relied upon to equal $ex(4X^2)$.

The Discharge Fallacy:

$$ex(\cdots Y \cdots |Y = X) = ex(\cdots X \cdots) \qquad \textbf{(NOT!)}$$

EXAMPLE 2, **The problem of the two sealed envelopes.** One contains a check for an unknown whole number of dollars, the other a check for twice or half as much. Offered a free choice, you pick one at random. What is wrong with the following argument for thinking you should have chosen the other?

> "Let X and Y be the values of the checks in the one and the other. As you think Y equally likely to be $.5X$ or $2X$, $ex(Y)$ will be $.5ex(.5X) + .5ex(2X) = 1.25ex(X)$, which is larger than $ex(X)$."

4.4 Median and Mean

Hydraulic Analogy. Let 'F' and 'S' mean heads on the first and second tosses of an ordinary coin. Suppose you stand to gain a dollar for each head. Then your net gain in the four possibilities for truth and falsity of F and S will be as shown at the left below.

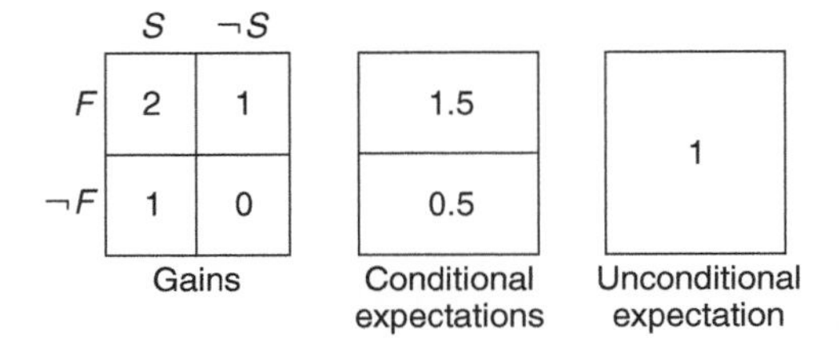

Think of that as a map of flooded walled fields in a plain, with the numbers indicating water depths in the four sections—e.g., the depth is $X = 2$ throughout the $F \wedge S$ region.[6] In the four regions, depths are values of X and areas are probabilities. To find your conditional expectation for X given F, remove the wall between the two sections of F so that the water reaches a single level in the two. That level will be $ex(X|F)$, which is 1.5 in the diagram. Similarly, removing the wall between the two sections of $\neg F$ shows that your conditional expectation for X given $\neg F$ is $ex(X|\neg F) = 0.5$. To find your unconditional expectation of gain, remove both walls so that the water reaches the same level throughout: $ex(X) = 1$.

But there is no mathematical reason for magnitudes X to have only a finite number of values, e.g., we might think of X as the birth weight in pounds of the next giant panda to be born in captivity—to no end of decimal places of accuracy, as if that meant something.[7] Nor is there any reason why X cannot have negative values, as in the case of temperature.

Balance. The following analogy is more easily adapted to the continuous case. On a weightless rigid beam, positions represent values of a magnitude X that might go negative as well as positive. Pick a zero, a unit, and a positive direction on the beam. Get a pound of modeling clay, and distribute it along the beam so that the weight of clay on each section represents your probability that the true value of X is in that section—say, like this:

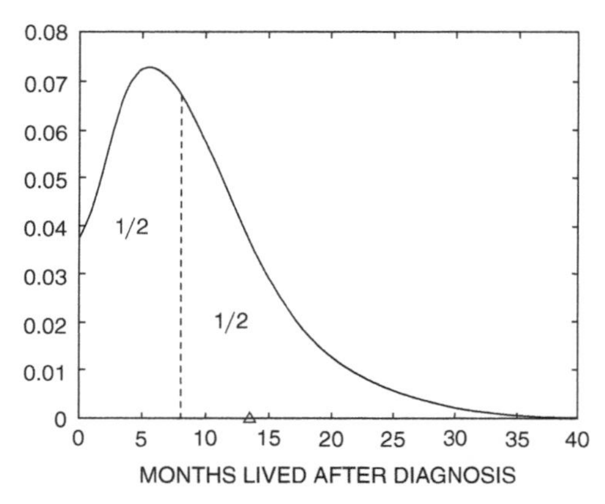

Locations on the beam are months lived after diagnosis; the weight of

[6] And therefore, $ex(X|F \wedge S) = 2$.

[7] It does not. The commonplace distinction between panda and ambient moisture, dirt, etc. is not drawn finely enough to let us take the remote decimal places seriously.

clay on the interval from 0 to m is the probability of still being alive in m months.

"The Median is not the Message."[8] "In 1982, I learned I was suffering from a rare and serious cancer. After surgery, I asked my doctor what the best technical literature on the cancer was. She told me ... that there was nothing really worth reading. I soon realized why she had offered that humane advice: my cancer is incurable, with a median mortality of eight months after discovery."

In terms of the beam analogy, here are the key definitions of the terms "median" and "mean"—the latter being a synonym for "expectation":

> The **median** is the point on the beam that divides the weight of clay in half: the probabilities are equal that the true value of X is represented by a point to the right and to the left of the median.
> The **mean** (= your expectation) is the point of support at which the beam would just balance.

Gould continues:

> The distribution of variation had to be right skewed, I reasoned. After all, the left of the distribution contains an irrevocable lower boundary of zero (since mesothelioma can only be identified at death or before). Thus there isn't much room for the distribution's lower (or left) half—it must be scrunched up between zero and eight months. But the upper (or right) half can extend out for years and years, even if nobody ultimately survives.

As this probability distribution is skewed (stretched out) to the right, the median is to the left of its mean; Gould's life expectancy is greater than 8 months. (The mean of 15 months suggested in the graph is my ignorant invention.)

The effect of skewness can be seen especially clearly in the case of discrete distributions like the following. Observe that if the right-hand weight is pushed further right the mean will follow, while the median remains between the second and third blocks.

[8] See Stephen Jay Gould, *Full House*, New York, 1996, sec. 4.

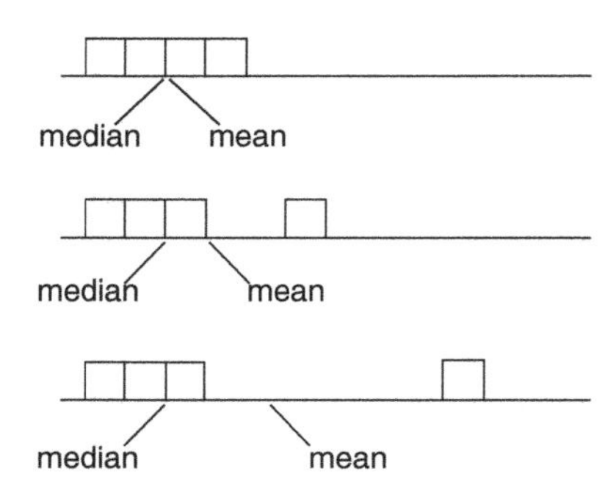

4.5 Variance

The **variance**, defined as[9]

$$varX = ex(X) - (exX)^2,$$

is one of the two commonly used measures of your uncertainty about the value of X. The other is the **"standard deviation"**: $\sigma(X) = \sqrt{varX}$. In terms of the physical analogies in sec. **4.4**, these measures of uncertainty correspond to the average spread of mass away from the mean. The obvious measure of this spread is your expectation $ex(|X - ex(X)|)$ of the absolute value of the difference, but the square, which is like the absolute value in being non-negative, is easier to work with, mathematically; thus, the variance. The move to the standard deviation counteracts the distortion in going from the absolute value to the square.

The definition of the variance can be simplified by writing the right-hand side as the expectation of the square minus the square of the expectation:[10]

(1) $\quad var(X) = ex([X - ex(X)]^2)$

Note that in case X is an indicator, $X = I_H$, we have

$$varI_H = pr(H)pr(\neg H) \leq \frac{1}{4}.$$

Proof. The inequality $pr(H)pr(\neg H) \leq \frac{1}{4}$ follows from the fact that $p(1-p)$ reaches its maximum value of 0.25 when $p = 0.5$, as shown in the graph:

[9] Here and below paradigm for 'ex^2X' is 'sin^2x' in trigonometry.
[10] *Proof.* $(X - ex(X))^2 = X^2 - 2Xex(X) + ex^2X$, where $ex(X)$ and its square are constants. By linearity of ex your expectation of this sum $= ex(X^2) - ex^2X$.

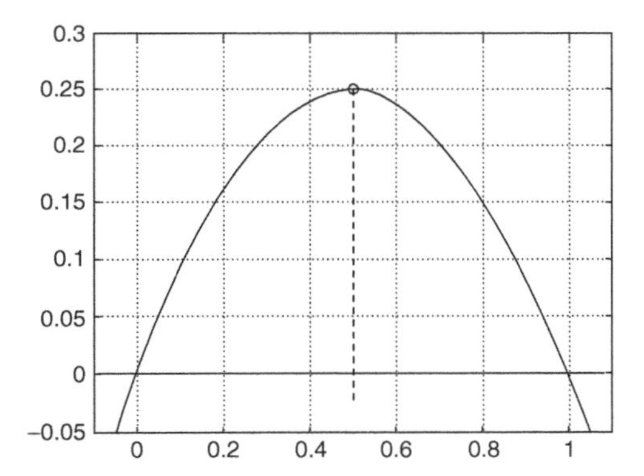

And the equality $var(I_H) = pr(H)pr(\neg H)$ follows from (1) because $(I_H)^2 = I_H$ (since I_H can only assume the values 0 and 1) and $ex(I_H) = pr(H)$.

In contrast to the linearity of expectations, variances are nonlinear: In place of $ex(aX + b) = a\,ex(X) + b$ we have

(2) $\quad var(aX + b) = a^2 var X.$

And in contrast to the unrestricted additivity of expectations, variances are additive only under special conditions—e.g., pairwise uncorrelation, which is Defined as follows:

> You see $X_1, \ldots, X_n$ as **pairwise uncorrelated** if and only if your $ex(X_i X_j) = ex(X_i)ex(X_j)$ whenever $i \neq j$.

Now to be a bit more explicit, this is what we want to prove:

(3) $\quad var \sum_{i=1}^{n} X_i = \sum_{i=i}^{n} var X_i$, given pairwise uncorrelation of $X_1, \ldots, X_n$.

Proof. Recall that by (1) and additivity of ex,

$$var(\sum_i X_i) = ex(\sum_i X_i)^2 - ex^2 \sum_i X_i = ex(\sum_i X_i)^2 - (ex \sum_i X_i)^2.$$

Multiply out squared sums and use ex additivity and pairwise noncorrelation to get $\sum_i ex(X)_i^2 + \sum_{i \neq j} ex(X)_i\, ex(X)_j - \sum_i ex^2 X_i - \sum_{i \neq j} ex(X)_i\, ex(X)_j$. Cancel identical terms with opposite signs and regroup to get $\sum_i (ex(X)_i^2 - ex^2 X_i)$. By (1), this $= \sum_i var X_i$.

The **covariance** of two R.V.'s is defined as

$$cov(X, Y) = ex[(X - ex(X))(Y - ex(Y))] = ex(XY) - ex(X)ex(Y)$$

so that the covariance of an R.V. with itself is the variance, $cov(X,X) = var(X)$. The **coefficient of correlation** between X and Y is defined as

$$\rho(X,Y) = \frac{cov(X,Y)}{\sigma(X)\sigma(Y)}$$

where σ is the positive square root of the variance. Noncorrelation, $\rho = 0$, is a consequence of independence, but it is a weaker property, which does not imply independence; see supplement **4.7.5** at the end of this chapter.

4.6 A Law of Large Numbers

Laws of large numbers tell you how to expect the "sample average" of the actual values of a number of random variables to behave as that number increases without bound:

$$\textbf{“Sample Average”},\ S_n = \frac{1}{n}\sum_{i=1}^{n} X_i$$

The basic question is how your expectations of sample averages of large finite numbers of random variables must behave. Here we show that, under certain rather broad conditions, your expectation of the squared difference $(S_{n'} - S_{n''})^2$ between a sample average and its prolongation will go to 0 as both their lengths, n' and n'', increase without bound:

If the random variables X_i all have the same finite expectations, $ex(X_i) = m$, variances $var(X_i) = \sigma^2$ and pairwise correlation coefficients, $\rho(X_i, X_j) = r$ where $i \neq j$, then for all positive n' and n'',

$$(4)\quad ex(S_{n'} - S_{n''})^2 = (\frac{1}{n'} - \frac{1}{n''})\sigma^2(1-r).$$

Proof.

$$(S_{n'} - S_{n''})^2 = (\frac{1}{n'}\sum_{i=1}^{n'} X_i)^2 + 2(\frac{1}{n'}\sum_{i=1}^{n'} X_i)(\frac{1}{n''}\sum_{i=1}^{n'} X_i) + (\frac{1}{n''}\sum_{i=1}^{n'} X_i)^2.$$

Multiplying this out we get terms involving $(X_i)^2$ and the rest involving X_iX_j with $i \neq j$. Since $m^2 + \sigma^2 = ex(X_i)^2$ and $m^2 + r\sigma^2 = ex(X_iX_j)$

with $i \neq j$, we can eliminate all of the X's and get formula (4) via some hideous algebra.[11]

4.7 Supplements

4.7.1 Markov's Inequality:

$$pr(X \geq \epsilon) \leq \frac{ex(X)}{\epsilon} \text{ if } \epsilon > 0 \text{ and } X \geq 0$$

This provides a simple way of getting an upper bound on your probabilities that non-negative random variables X for which you have finite expectations are at least ϵ away from 0. *Proof of the inequality.* If $Y = \{{\epsilon \text{ when } X \geq \epsilon \atop 0 \text{ when } X < \epsilon}\}$ then by the law of total expectation, $ex(Y) = \epsilon \cdot pr(X \geq \epsilon) + 0 \cdot pr(X < \epsilon) = \epsilon \cdot pr(X \geq \epsilon)$, so that $pr(X \geq \epsilon) = \frac{ex(Y)}{\epsilon}$. Now since $Y \leq X$ both when $X \geq \epsilon$ and when $X < \epsilon$ we have $ex(Y) \leq ex(X)$ and, so, Markov's inequality.

4.7.2 Chebyshev's Inequality:

$$pr(|X - ex(X)| \geq \epsilon) \leq \frac{var(X)}{\epsilon^2} \text{ if } \epsilon > 0$$

Proof. Since squares are never negative, the Markov inequality implies that $pr(X^2 \geq \epsilon) \leq ex(X^2)/\epsilon$ if $\epsilon > 0$; and from this, since $|X| \geq \epsilon$ if and only if $X^2 \geq \epsilon^2$, we have $pr(|X| \geq \epsilon) \leq ex(X^2)/\epsilon^2$ if $\epsilon > 0$. From this, with '$X - ex(X)$' for 'X', we have Chebyshev's inequality.

4.7.3 Many Pairwise Uncorrelated R.V.'s

An updated version of the oldest law of large numbers (1713, James Bernoulli's) applies to random variables $X_1, \ldots, X_n$ which you see as pairwise uncorrelated and as having a common, finite upper bound, b, on their variances. Here there will be a finite upper bound $b/\epsilon^2 n$ on your probability that the sample average will differ by more than ϵ from your expectation of that average. The bound b on the X's is independent of the sample size, n, so that for large enough samples your probability that the error $|S_n - ex(S_n)|$ in your expectation is ϵ or more gets as small as you please.

[11] See Bruno de Finetti, *Theory of Probability*, vol. 2, pp. 215–216. For another treatment, see pp. 84–85 of de Finetti's 'Foresight' in Henry Kyburg and Howard Smokler (eds.), *Studies in Subjective Probability*, Huntington, N.Y., Krieger, 1980.

For pairwise uncorrelated R.V.'s X_i whose variances are all $\leq b$:

(5) For any $\epsilon > 0$: $\quad pr(|S_n - ex(S_n)| \geq \epsilon) \leq \dfrac{b}{n\epsilon^2}$

This places an upper limit of $\frac{b}{n\epsilon^2}$ on your probability that the sample average differs from your expectation of it by ϵ or more.

Proof. By (3), $var(S_n) = \dfrac{1}{n^2}\sum_{i=1}^{n} var(X_i)$. Since each $var(X_i)$ is $\leq b$, this is $\leq \dfrac{1}{n^2}(nb) = \dfrac{b}{n}$. Setting $X = S_n$ in Chebyshev's inequality, we have (5).

Note that if the X_i are indicators of propositions H_i, sample averages S_n are relative frequencies R_n of truth:

$$R_n = \frac{1}{n}\sum_{i=1}^{n} I_{H_i} = \frac{1}{n}\text{ (the number of truths among } H_1, \ldots, H_n)$$

4.7.4 Corollary: Bernoulli Trials

These are sequences of propositions H_i (say, "head on the i'th toss") where for some p ($= 1/2$ for coin-tossing) your probability for truth of any n distinct H's is p^n.

Corollary. Suppose that, as in the case of Bernoulli trials, for all distinct i and j from 1 to n : $pr(H_i) = pr(H_1)$ and $pr(H_i \wedge H_j) = pr^2(H_1)$. Then

(6) For any $\epsilon > 0$, $\quad pr(|R_n - pr(H_1)| \geq \epsilon) \leq \dfrac{1}{4n\epsilon^2}$.

This places a limit of $1/4n\epsilon^2$ on your pr for a deviation of ϵ or more between your $pr(H_i)$ and the actual relative frequency of truths in n trials.

Thus, in $n = 25{,}000$ tosses of a fair coin, your probability that the success rate will be within $\epsilon = .01$ of 0.50 will fall short of 1 by at most $1/4n\epsilon^2 = 10\%$.

Proof. Since $ex(X)_i = ex(I_{H_i}) = pr(H_i) = p$, linearity of ex says that $ex(S_n) = \frac{1}{n} \cdot np = p$. Now by (2) we can set $b = \frac{1}{4}$ in (5).

4.7.5 Noncorrelation and Independence

Random variables X, Y are said to be **uncorrelated** from your point of view when your expectation of their product is simply the product of your separate expectations (sec. **4.5**). In the special case where X and Y are the indicators of propositions, noncorrelation reduces to independence:

> $ex(XY) = ex(X)ex(Y)$ iff: $pr(X = x \wedge Y = y) = pr(X = x)pr(Y = y)$ for all four pairs (x, y) of zeroes and ones, or even for the single pair $X = Y = 1$.

But in general the uncorrelated pairs of random variables are not simply the independent ones, as the following extreme example demonstrates.

EXAMPLE, **Noncorrelation in spite of deterministic dependency.**[12] You have probability 1/4 for each of $X = -2, -1, 1, 2$; and $Y = X^2$, so that you have probability 1/2 for each possible value of $Y (= 1$ or $4)$, and probability 1/4 for each conjunction $X = x \wedge Y = x^2$ with $x = -2, -1, 1, 2$. Then you see X and Y as uncorrelated, for $ex(X) = 0$ and $ex(XY) = \frac{-8-1+1+8}{4} = 0 = ex(X)ex(Y)$. Yet, $pr(X = 1 \wedge Y = 1) = \frac{1}{4} \neq pr(X = 1)pr(Y = 1) = (\frac{1}{4})(\frac{1}{2})$.

Then noncorrelation need not imply independence. But independence does always imply noncorrelation. In particular, in cases where X and Y take only finitely many values, we have:[13]

> If $pr(X = x \wedge Y = y) = pr(X = x) \cdot pr(Y = y)$ for all values x of X and y of Y, then $ex(XY) = ex(X)ex(Y)$.

[12] From William Feller, *An Introduction to Probability Theory and its Applications.* vol. 1, 2nd ed., 1957, p. 222, example (a).

[13] *Proof.* By formula (1) in sec. **4.1**, $ex(XY) = \sum_{i,j} x_i y_j \, pr(X = x_i \wedge Y = y_j)$, or, by independence, $\sum_{i,j} x_i y_j \, pr(X = x_i) \, pr(Y = y_j)$. By the same formula, $ex(X)ex(Y) = \{\sum_i x_i \, pr(X = x_i)\}\{\sum_j y_j \, pr(Y = y_j)\}$, which $= \sum_{i,j} x_i y_j \, pr(X = x_i) pr(Y = y_j)$ again.

5
Updating on Statistics

5.1 Where Do Probabilities Come from?

Your "subjective" probability is not something fetched out of the sky on a whim;[1] it is what your actual judgment *should* be, in view of your information to date and of your sense of other people's information, even if you do not regard it as a judgment that everyone must share on pain of being wrong in one sense or another.

But of course you are not always clear about what your judgment is, or should be. The most important questions in the theory of probability concern ways and means of constructing reasonably satisfactory probability assignments to fit your present state of mind. (Think: trying on shoes.) For this, there is no overarching algorithm. Here we examine two answers to these questions that were floated by Bruno de Finetti in the decade from (roughly) 1928 to 1938. The second of them, "Exchangeability",[2] postulates a definite sort of initial probabilistic state of mind, which is then updated by conditioning on statistical data. The first ("Minimalism") is more primitive: The input probability assignment will have large gaps, and the output will not arise via conditioning.

5.1.1 Probabilities from Statistics: Minimalism

Statistical data are a prime determinant of subjective probabilities; that is the truth in frequentism. But that truth must be understood in the light of certain features of judgmental probabilizing. One such feature

[1] In common usage there is some such suggestion, and for that reason I would prefer to speak of 'judgmental' probability. But in the technical literature the term 'subjective' is well established—and this is in part because of the lifelong pleasure de Finetti found in being seen to give the finger to the establishment.

[2] And, more generally, "partial" exchangeability (see **5.2** below), of which simple exchangeability, which got named earlier, is a special case.

that can be of importance is persistence, as you learn the relative frequency of truths in a sequence of propositions, of your judgment that they all have the same probability. That is a condition under which the following little theorem of the probability calculus can be used to generate probability judgments.[3]

> **Law of Little Numbers.** In a finite sequence of propositions that you view as equiprobable, if you are sure that the relative frequency of truths is p, then your probability for each is p.

Then, if judging a sequence of propositions to be equiprobable, you learn the relative frequency of truths *in a way that does not change your judgment of equiprobability,* your probability for each proposition will agree with the relative frequency.[4]

The law of little numbers can be generalized to random variables:

> **Law of Short Run Averages.** In a finite sequence of random variables for which your expectations are equal, if you know only their arithmetical mean, then that is your expectation of each.

Proof. By linearity of ex, if $ex(X_i) = p$ for $i = 1, \ldots, n$, $p = \frac{1}{n}\sum_{i=1}^{n} ex(X_i)$.

Then, if while requiring your final expectations for a sequence of magnitudes to be equal, you learn their mean value in a way that does not lead you to change that requirement, your expectation of each will agree with that mean.[5]

EXAMPLE, **Guessing weight**. Needing to estimate the weight of someone on the other side of a chain-link fence, you select ten people on

[3] See Jeffrey (1992), pp. 59–64. The name "Law of Little Numbers" is a joke, but I know of no generally understood name for the theorem. That theorem, like the next (the "Law of Short Run Averages", another joke) is quite trivial; both are immediate consequences of the linearity of the expectation operator. Chapter 2 of de Finetti (1937) is devoted to them. In chapter 3 he goes on to a mathematically deeper way of understanding the truth in frequentism, in terms of "exchangeability" of random variables (sec. **1.2**, below).

[4] To appreciate the importance of the italicized caveat, note that if you learn the relative frequency of truths by learning which propositions in the sequence are true, and which false, and as you form your probabilities for those propositions you remember what you have learned, then those probabilities will be zeros and ones instead of averages of those zeros and ones.

[5] If you learn the individual values and calculate the mean as their average without forgetting the various values, you have violated the caveat (unless it happens that all the values were the same), for what you learned will have shown you that they are not equal.

your side whom you estimate to have the same weight as that eleventh, persuade them to congregate on a platform scale, and read their total weight. If the scale reads 1080 lb., your estimate of the eleventh person's weight will be 108 lb.—if nothing in that process has made you revise your judgment that the eleven weights are equal.[6]

This is a frequentism in which judgmental probabilities are seen as judgmental *expectations* of frequencies, and in which the Law of Little Numbers guides the recycling of observed frequencies as probabilities of unobserved instances. It is to be distinguished both from the intelligible but untenable finite frequentism that simply identifies probabilities with actual frequencies (generally, unknown) when there are only finitely many instances overall, and from the unintelligible long-run frequentism that would see the observed instances as a finite fragment of an infinite sequence in which the infinitely long run inflates expectations into certainties that sweep judgmental probabilities under the endless carpet.

5.1.2 Probabilities from Statistics: Exchangeability[7]

On the hypotheses of (a) equiprobability and (b) certainty that the relative frequency of truths is r, the Law of Little Numbers identified the probability as r. Stronger conclusions follow from the stronger hypothesis of

> **Exchangeability:** You regard propositions $H_1, \ldots, H_n$ as exchangeable when, for any particular t of them, your probability that they are all true and the other $f = n - t$ false depends only on the numbers t and f.[8] Propositions are essentially indicators, random variables taking only the values 0 (falsity) and 1 (truth).

[6] Note that turning statistics into probabilities or expectations in this way requires neither conditioning nor Bayes's theorem, nor does it require you to have formed particular judgmental probabilities for the propositions or particular estimates for the random variables prior to learning the relative frequency or mean.

[7] Bruno de Finetti, *Theory of Probability*, vol. 2, Wiley, 1975, pp. 211–224.

[8] This comes to the same thing as invariance of your probabilities for Boolean compounds of finite numbers of the H_i under all finite permutations of the positive integers, e.g., $P(H_1 \wedge (H_2 \vee \neg H_3)) = P(H_{100} \wedge (H_2 \vee \neg H_7))$.

For random variables $X_1, \ldots, X_n$ more generally, exchangeability means invariance of $pr(X_1 < r_1) \wedge \ldots \wedge (X_n < r_n)$ for each n-tuple of real numbers $r_1, \ldots, r_n$ and each permutation of the X's that appear in the inequalities.

Here, again, as in sec. **5.1.1**, probabilities will be seen to come from statistics—this time, by ordinary conditioning, under the proviso that your prior probability assignment was exchangeable.

Exchangeability turns out to be a strong but reasonably common condition under which the hypotheses of the law of large numbers in sec. **4.6** are met.

5.1.3 Exchangeability: Urn Examples

The following illustration in the case $n = 3$ will serve to introduce the basic ideas.

EXAMPLE 1, **Three draws from an urn of unknown composition.** The urn contains balls that are red (**r**) or green (**g**), in an unknown proportion. If you know the proportion, that would determine your probability for red. But you don't. Still, you may think that (say) red is the more likely color. How likely? Well, here is one precise probabilistic state of mind you might be in: Your probabilities for drawing a total of 0, 1, 2, 3 red balls might be $p_0 = 0.1$, $p_1 = 0.2$, $p_2 = 0.3$, $p_3 = 0.4$. *And when there is more than one possible order in which a certain number of heads can show up, you will divide your probability among them equally.* In other words: *You see the propositions* H_i (= red on the i'th trial) *as exchangeable.*

There are eight possibilities about the outcomes of all three trials:

rrr | **rrg rgr grr** | **rgg grg ggr** | **ggg**

The vertical lines divide clumps of possibilities in which the number of reds is 3, 2, 1, 0. In Table 1, the possibilities are listed under 'h' (for 'history') and again under the H's, where the 0's and 1's are truth values $(1 = true, 0 = false)$ as they would be in the eight possible histories. The value of the random variable T_3 ("tally" of the first 3 trials) at the right is the number of r's in the actual history, the number of 1's under the H's in that row. The entries in the rightmost column are the values of p_t taken from example 1.

h	H_1	H_2	H_3	$pr\{h\}$	t	$pr(T_3 = t)$
rrr	1	1	1	p_3	3	0.4
rrg	1	1	0	$\frac{p_2}{3}$		
rgr	1	0	1	$\frac{p_2}{3}$	2	0.3
grr	0	1	1	$\frac{p_2}{3}$		
rgg	1	0	0	$\frac{p_1}{3}$		
grg	0	1	0	$\frac{p_1}{3}$	1	0.2
ggr	0	0	1	$\frac{p_1}{3}$		
ggg	0	0	0	p_0	0	0.1

Table 1. Three Exchangeable Trials

We can think of propositions as identified by the sets of histories in which they would be true. Examples: $H_1 = \{\texttt{rrr}, \texttt{rrg}, \texttt{rgr}, \texttt{rgg}\}$ and $H_1 \wedge H_2 \wedge H_3 = \{\texttt{rrr}\}$. And by additivity, your probability for "same outcome on the first two trials" is $pr\{\texttt{rrr}, \texttt{rrg}, \texttt{ggr}, \texttt{ggg}\} = p_1 + \frac{1}{3}p_2 + \frac{1}{3}p_1 + p_0 = .4 + \frac{.3}{3} + \frac{.2}{3} + 0.1 = \frac{2}{3}$.

Evidently, exchangeability undoes the dreaded "combinatorial explosion":

> **Combinatorial Implosion.** Your probabilities for all $2^{(2^n)}$ truthfunctional compounds of n exchangeable H_i are determined by your probabilities for a particular n of them—namely, by all but one of your probabilities $p_0, p_1, \ldots, p_n$ for the propositions saying that the number of truths among the H's is $0, 1, \ldots, n$.[9]

The following important corollary of the combinatorial implosion is illustrated and proved below. (The R.V. T_m is the tally of the first m trials.)

> **The Rule of Succession.**[10] If $H_1, \ldots, H_n, H_{n+1}$ are exchangeable for you and $\{h\}$ is expressible as a conjunction of t of the first n H's with the denials of the other $f = n - t$, then
> $$pr(H_{n+1}|\{h\}) = \frac{t+1}{n+1} \times \frac{pr(T_{n+1} = t+1)}{pr(T_n = n)} = \frac{t+1}{n+2+*},$$
> where the correction term, $*$, is $(f+1)\,(\frac{pr(T_{n+1}=t+1)}{pr(T_n=n)} - 1)$.

EXAMPLE 2, **Uniform distribution:** $pr(T_{n+1} = t) = \frac{1}{n+2}$, where t

[9] The probability of the missing one will be 1 minus the sum of the other n.

[10] The best overall account is by Sandy Zabell, "The Rule of Succession" *Erkenntnis* **31** (1989) 283–321.

ranges from 0 to $n+1$. In this ("Bayes-Laplace-Johnson-Carnap")[11] case the correction term $*$ in the rule of succession vanishes, and the rule reduces to

$$pr(H_{n+1}|\{h\}) = pr(H_{n+1}|T_n = t) = \frac{t+1}{n+2}.$$

EXAMPLE 3, **The ninth draw.** If you adopt the uniform distribution, your probability for red next, given 6 red and 2 green on the first 8 trials, will be $pr(H_9|T_8 = 6) = pr(H_9|\{h\})$, where h is any history in the set that represents the proposition $T_8 = 6$. Then your $pr(H_9|T_8 = 6) = \frac{7}{10}$.

Proof of the rule of succession.

(a) $pr(H_{n+1}|C_k^n) = \dfrac{pr(H_{n+1} \wedge C_k^n)}{pr(C_k^n)} = \dfrac{pr(H_{t+1}^{n+1})/\binom{n+1}{t+1}}{pr(H_t^n)/\binom{n}{t}} =$

$\dfrac{(t+1)pr(H_{t+1}^{n+1})}{(n+1)pr(H_t^n)}$ by the quotient rule, implosion, and the fact that $\binom{n}{t}/\binom{n+1}{t+1} = (t+1)/(n+1)$;

(b) $pr(C_k^n) = pr(\neg H_{n+1} \wedge C_k^n) + pr(H_{n+1} \wedge C_k^n)$;

(c) $\dfrac{pr(H_t^n)}{\binom{n}{t}} = \dfrac{pr(H_t^{n+1})}{\binom{n+1}{t}} + \dfrac{pr(H_{t+1}^{n+1})}{\binom{n+1}{t+1}}$;

(d) $\dfrac{\binom{n}{t}}{\binom{n+1}{t}} = \dfrac{f+1}{n+1}$ and $\dfrac{\binom{n}{t}}{\binom{n+1}{t+1}} = \frac{t+1}{n+1}$;

(e) $pr(H_t^n) = \dfrac{f+1}{n+1}pr(H_t^{n+1}) + \dfrac{t+1}{n+1}pr(H_{t+1}^{n+1})$;

(f) $\dfrac{(t+1)pr(H_{t+1}^{n+1})}{(n+1)pr(H_t^n)} = \dfrac{(t+1)pr(H_{t+1}^{n+1})}{(f+1)pr(H_t^{n+1}) + (t+1)} \left(\dfrac{pr(H_t^{n+1})}{pr(H_{t+1}^{n+1})} - 1\right)$;

(g) the right-hand side of (f) $= \dfrac{t+1}{n+2+(f+1)} \left(\dfrac{pr(H_t^{n+1})}{pr(H_{t+1}^{n+1})} - 1\right)$.

5.1.4 Supplements

1. *Law of Large Numbers.* Show that exchangeable random variables satisfy the three conditions in the law of **4.6.**
2. *A Pólya Urn Model.* At stage 0, an urn contains two balls, one red and one green. At each later stage a ball is drawn with *double replacement:* It is replaced along with another ball of its color. Given that black was drawn on the first t trials and white on the next $f = n - t$ trials, prove that the conditional odds on black's

[11] For the history underlying this terminology, see Zabell's paper, cited above.

being drawn on the $n+1$'st trial, are $\frac{t+1}{f+1}$, exactly as in the urn examples of sec. **5.1.3** above.

3. *Exchangeability is preserved under mixing.* Prove that if $H_1, \ldots, H_n$ are exchangeable relative to each of $pr_1, \ldots, pr_m$ then they are exchangeable relative to $w_1 pr_1 + \ldots + w_m pr_m$ if the w's are non-negative and sum to 1.
4. *Independence is not preserved under mixing.* Prove this, using a simple counterexample where $m = 2$ and $w_1 = w_2$.
5. *Independence, Conditional Independence, Exchangeability.* Suppose you are sure that the balls in an urn are either 90% red or 10% red, and you regard those two hypotheses as equiprobable. If balls are drawn with replacement, then conditionally on each hypothesis the propositions $H_1, H_2, \ldots$ that the first, second, etc., balls drawn will be red are independent and equiprobable. Is that also so unconditionally?
   (a) Are the H's unconditionally equiprobable? If so, what is their probability? If not, why not?
   (b) Are they unconditionally independent? Would your probability for red next be the same as it was initially after drawing one ball, which is red?
   (c) Are they exchangeable?

Diaconis and Freedman on de Finetti's Generalizations of Exchangeability[12]

De Finetti wrote about partial exchangeability over a period of half a century.[13] These treatments are rich sources of ideas, which take many readings to digest. Here we give examples of partial exchangeability that we understand well enough to put into crisp mathematical terms. All results in Sections **2–5** involve random quantities taking only two values. We follow de Finetti by giving results for finite as well as infinite sequences. In **5.3** we review exchangeability. In **5.4** we present examples of partially exchangeable sequences—2×2 tables and Markov chains—and give general definitions. In **5.5** a finite form of de Finetti's theorem

[12] This is an adaptation of the chapter (#11) by Persi Diaconis and David Freedman in *Studies in Inductive Logic and Probability*, vol. 2, Richard Jeffrey (ed.), University of California Press, 1980, which is reproduced here with everybody's kind permission. A bit of startlingly technical material in the original publication (see **5.5** and **5.6**) has been retained here, suitably labeled.

[13] Beginning with de Finetti (1938).

is presented. **5.6** gives some infinite versions of de Finetti's theorem and a counterexample which shows that the straightforward generalization from the exchangeable case sketched by de Finetti is not possible. The last section contains comments about the practical implications of partial exchangeability. We also discuss the closely related work of P. Martin-Löf on repetitive structures.

5.2 Exchangeability Itself

We begin with the case of exchangeability. Consider the following experiment: A coin is spun on a table. We will denote heads by 1 and tails by 0. The experiment is to be repeated ten times. You believe you can assign probabilities to the $2^{10} = 1024$ possible outcomes by looking at the coin and thinking about what you know. There are *no* a priori restrictions on the probabilities assigned save that they are nonnegative and sum to 1. Even with so simple a problem, assigning over a thousand probabilities is not a simple task. De Finetti has called attention to exchangeability as a possible simplifying assumption. With an exchangeable probability, two sequences of length 10, with the same number of ones, are assigned the same probability. Thus, the probability assignment is symmetric, or invariant under changes in order. Another way to say this is that only the number of ones in the ten trials matters, not the location of the ones. If believed, the symmetry assumption reduces the number of probabilities to be assigned from 1024 to 11—the probability of no ones through the probability of ten ones.

It is useful to single out certain extreme exchangeable probability assignments. Though it is unrealistic, we might be sure there will be exactly one head in the ten spins. The assumption of exchangeability forces each of the ten possible sequences with a single one in it to be equally likely. The distribution is just like the outcome of ten draws without replacement from an urn with one ball marked 1 and nine balls marked 0—the ball marked 1 could be drawn at any stage and must be drawn at some time. There are 11 such extremal urns corresponding to sure knowledge of exactly i heads in the ten draws where i is a fixed number between zero and ten. The first form of de Finetti's theorem follows from these observations:

> **(1) Finite form of de Finetti's theorem:** Every exchangeable probability assignment on sequences of length N is a unique mixture of draws WITHOUT replacement from the $N + 1$ extremal urns.[14]

[14] This form of the result was given by de Finetti (1938) and by many other writers on the subject. Diaconis (1977), pp. 271–281, is an accessible treatment.

By a mixture of urns we simply mean a probability assignment over the $N+1$ possible urns. Any exchangeable assignment on sequences of length N can be realized by first choosing an urn and then drawing without replacement until the urn is empty.

The extremal urns, representing certain knowledge of the total number of ones, seem like unnatural probability assignments in most cases. While such situations arise (for example, when drawing a sample from a finite population), the more usual situation is that of the coin. While we are only considering ten spins, in principle it seems possible to extend the ten to an arbitrarily large number. In this case there is a stronger form of (1) which restricts the probability assignment within the class of exchangeable assignments.

> **(2) Infinite form of de Finetti's theorem:** Every exchangeable probability assignment pr that can be indefinitely extended and remain exchangeable is a unique mixture μ of draws WITH replacement from a possibly infinite array of extremal urns.

Again the mixture is defined by a probability assignment to the extremal urns. Let the random variable p be the proportion of red balls in the urn from which the drawing will be made. A simple example where the array is a continuum: For each sub-interval I of the real numbers from 0 to 1, $\mu(p \in I) =$ the length of I. This is the μ that is uniquely determined when pr is the uniform distribution (**5.1.3,** example 2). If $\sum_{p \in C} \mu(p) = 1$ for a countable set C of real numbers from 0 to 1, the formula is

(3) $$pr(j \text{ ones in } k \text{ trials}) = \binom{k}{j} \sum_{p \in C} p^k (1-p)^j \mu(p)$$

In the general case $C = [0,1]$ and the sum becomes an integral.[15] This holds for every k with the same μ. Spelled out in English, (3) and its general case become (2). We have more to say about this interpretation of exchangeability as expressing ignorance of the true value of p in Remark 1 of **5.6.**

Not all exchangeable measures on sequences of length k can be extended to exchangeable sequences of length $n > k$. For example, sampling without replacement from an urn with k balls in it cannot be

[15] $pr(j \text{ ones in } k \text{ trials}) = \binom{k}{j} \int_{p \in [0,1]} p^k (1-p)^j d\mu(p)$.

extended to $k+1$ trials.[16] The requirement that an exchangeable sequence of length k be infinitely extendable seems out of keeping with de Finetti's general program of restricting attention to finite samples. An appropriate finite version of (3) is given by Diaconis and Freedman (1978), where we show that if an exchangeable probability on sequences of length k is extendable to an exchangeable probability on sequences of length $n > k$, then (3) almost holds in the sense that there is a measure μ such that for any set $A \subseteq \{0, 1, 2, \ldots, k\}$,

$$\textbf{(4)} \qquad \left| pr \left\{ \begin{matrix} \text{number of 1's in} \\ k \text{ trials is in } A \end{matrix} \right\} - B_\mu \left\{ \begin{matrix} \text{number of 1's in} \\ k \text{ trials is in } A \end{matrix} \right\} \right| \leq \frac{2k}{n}$$

uniformly in n, k, and A—where B_μ is the assignment determined by the mixture μ of drawings with replacement from urns with various proportions p of 1's.[17]

For example, it is easy to imagine the spins of the coin as the first ten spins in a series of 1000 spins. This yields a bound of .02 on the right hand side of (4).

Both of the results (3) and (4) imply that for many practical purposes, instead of specifying a probability assignment on the number of ones in n trials, it is equivalent to specify a prior measure μ on the unit interval. Much of de Finetti's discussion in his papers on partial exchangeability is devoted to reasonable choices of μ.[18] We will not discuss the choice of a prior further but rather restrict attention to generalizations of (2), (3), and (4) to partially exchangeable probability assignments.

5.3 Two Species of Partial Exchangeability

In many situations exchangeability is not believable or does not permit incorporation of other relevant data. Here are two examples which will be discussed further in **5.4** and **5.5.**

[16] Necessary and suffcient conditions for extension are found in de Finetti (1969), Crisma (1971) and Diaconis (1977).

[17] Hairy-style (4): $|pr\left\{ \begin{matrix} \text{number of 1's in} \\ k \text{ trials is in } A \end{matrix} \right\} - \sum_{j \in A} \binom{k}{j} \int_{p \in [0,1]} p^j (1-p)^{k-1} d\mu(p)| \leq \frac{2k}{n}$.

[18] The paper by A. Bruno (1964) "On the notion of partial exchangeability", *Giorn. 1st. It. Attuari* **27**, 174–196, translated in chapter 10 of de Finetti (1974), is also devoted to this important problem.

5.3.1 2 × 2 Tables

Consider zero/one outcomes in a medical experiment with n subjects. We are told each subject's sex and if each subject was given a treatment or was in a control group. In some cases, a reasonable symmetry assumption is the following kind of partial exchangeability: Regard all the treated males as exchangeable with one another but not with the subjects in the other three categories; likewise for the other categories. Thus, two sequences of zeros and ones of length n which had the same number of ones in each of the four categories would be assigned the same probability. For example, if $n = 10$, each of the three sequences below must be assigned the same probability.

Trial	1	2	3	4	5	6	7	8	9	10
Sex	M	M	M	F	F	M	M	F	F	F
Treatment/Control	T	T	T	T	T	C	C	C	C	C
1	1	0	1	0	1	1	1	0	1	0
2	0	1	1	0	1	1	1	0	0	1
3	0	1	1	1	0	1	1	1	0	0

In this example, there are three treated males, two treated females, two control males, and three control females. The data from each of the three sequences can be summarized in a 2 × 2 table which records the number of one outcomes in each group. Each of the three sequences leads to the same table:

(5)

	T	C
M	2	2
F	1	1

5.3.2 Markov Dependency

Consider an experiment in which a thumbtack is placed on the floor and given an energetic flick with the fingers. We record a one if the tack lands point upward and a zero if it lands point to the floor. For simplicity suppose the tack starts point to the floor. If after each trial the tack were reset to be point to the floor, exchangeability might be a tenable assumption. If each flick of the tack was given from the position in which the tack just landed, then the result of each trial may depend on the result of the previous trial. For example, there is some chance the tack will slide across the floor without ever turning. It does not seem reasonable to think of a trial depending on what happened two or more trials before. A natural notion of symmetry here is to say that if two sequences of zeros and ones of length n which both begin with zero have the same number of transitions: zero to zero, zero to one, one to zero, and one to one, they should both be assigned the same probability. For example, if $n = 10$, any of the following three sequences would be assigned the same probability.

Trial	1	2	3	4	5	6	7	8	9	10
1	0	1	0	1	1	0	0	1	0	1
2	0	1	0	1	0	0	1	1	0	1
3	0	0	1	0	1	0	1	0	1	1

Each sequence begins with a zero and has the same transition matrix

(6)

		to	
		0	1
from	0	1	4
	1	3	1

It turns out that there are 16 different sequences starting with zero that have this transition matrix.

A general definition of partial exchangeability that includes these examples involves the notion of a statistic: A function from the sequences of length n into a set X. A probability assignment P on sequences of length n is partially exchangeable for a statistic T if

$$T(x) = T(y) \text{ implies } P(x) = P(y)$$

where x and y are sequences of length n. Freedman (1962) and Lauritzen (1974) have said that P was summarized by T in this situation. In the case of exchangeability the statistic T is the number of ones. Two sequences with the same number of ones get assigned the same probability by an exchangeable probability. This definition is equivalent to the usual one of permutation invariance. In the case of 2×2 tables the statistic T is the 2×2 table (5). In the Markov example the statistic T

is the matrix of transition counts (6) along with the outcome of the first trial.

Such examples can easily be combined and extended. For instance, one can consider a two-stage Markov dependence, with additional information given such as which experimenter reported the trial as well as the time of day. De Finetti (1938, 1974, Section 9.6.2), Martin-Löf (1970, 1974), and Diaconis and Freedman (1978a,b,c) give further examples. In remark 4 of sec. **6** we discuss a more general definition of partial exchangeability.

5.4 Finite Forms of de Finetti's Theorem on Partial Exchangeability

There is a simple analog of Theorem (1) for partially exchangeable sequences of length n. We will write $\{0,1\}^n$ for the set of sequences of zeros and ones of length n. Let $T : \{0,1\}^n \to X$ be a statistic taking values $t_1, t_2, \ldots, t_k$. Let $S_i = \{x \in 2^n : T(x) = t_i\}$ and suppose S_i contains n_i elements. Let pr_i be the probability assignment on 2^n which picks a sequence $x \in 2^n$ by choosing an x from S_i uniformly—e.g., with probability $1/n_i$. Then pr_i is partially exchangeable with respect to T. In terms of these definitions we now state

> **(7)** A finite form of de Finetti's theorem on partial exchangeability. Every probability assignment pr on $\{0,1\}^n$ which is partially exchangeable with respect to T is a unique mixture of the extreme measures pr_i. The mixing weights are $wi = pr\{x : T(x) = t_i\}$.[19]

Theorem (7) seems trivial but in practice a more explicit description of the extreme measures P_i—like the urns in (1)—can be difficult. We now explore this for the examples of Section 3.

(8) Example—2 $\times$ 2 tables

Suppose we know there are: a treated males, b untreated males, c treated females, and d untreated females with $a + b + c + d = n$. The sufficient statistic is the 2 $\times$ 2 table with entries which are the number

[19] In the language of convex sets, the set of partially exchangeable probabilities with respect to T forms a simplex with extreme points pr_i.

of ones in each of the four groups.

$$\begin{array}{cc} & \begin{array}{cc} T & U \end{array} \\ \begin{array}{c} M \\ F \end{array} & \begin{pmatrix} i & j \\ k & l \end{pmatrix} \end{array} \quad \text{where} \quad \begin{array}{ll} 0 \le i \le a, & 0 \le j \le b \\ 0 \le k \le c, & 0 \le l \le d. \end{array}$$

There are $(a+1) \times (b+1) \times (c+1) \times (d+1)$ possible values of the matrix. An extreme partially exchangeable probability can be thought of as follows: Fix a possible matrix. Make up four urns. In the first urn, labeled TM (for treated males), put i balls marked one and $a - i$ balls marked zero. Similarly, construct urns labeled UM, TF, and UF. To generate a sequence of length n given the labels (Sex and Treated/Untreated) draw without replacement from the appropriate urns. This scheme generates all sequences $x \in 2^n$ with the given matrix $\binom{i\ j}{k\ l}$ equiprobably. Theorem (7) says that any partially exchangeable probability assignment on 2^n is a unique mixture of such urn schemes. If a, b, c, and d all tend to infinity, the binomial approximation to the hypergeometric distribution will lead to the appropriate infinite version of de Finetti's theorem as stated in Section 5.

(9) Example—Markov chains

For simplicity, assume we have a sequence of length $n+1$ that begins with a zero. The suffcient statistic is $T = \binom{t_{00}\ t_{01}}{t_{10}\ t_{11}}$ where t_{ij} is the number of i to j transitions. A counting argument shows that there are $\binom{n}{2} + 1$ different values of T possible. Here is an urn model which generates all sequences of length $n+1$ with a fixed transition matrix T equiprobably. Form two urns U_0 and U_1 as follows: Put t_{ij} balls marked j into urns U_i. It is now necessary to make an adjustment to make sure the process doesn't run into trouble. There are two cases possible:

Case 1. If $t_{01} = t_{10}$ remove a zero from U_1.
Case 2. If $t_{01} = t_{10} + 1$, remove a one from U_0.

To generate a sequence of length $n+1$, let $X_1 = 0$. Let X_2 be the result of a draw from U_x; and, in general, let X_i be the result of a draw without replacement from urn $U_{x_{i-1}}$. If the sequence generated ever forces a draw from an empty urn, make a forced transition to the other urn. The adjustment made above guarantees that such a forced jump can only be made once from either urn and that the process generates all sequences of length $n+1$ that start with zero and have transition matrix T with the same probability. Theorem (7) says that every probability assignment on sequences of length $n+1$ which is partially exchangeable for the transition matrix T is a unique mixture of the $\binom{n}{2}+1$ different urn processes described above. Again, the binomial approximation to the hypergeometric will lead to an infinite form of de Finetti's theorem in certain cases. This is further discussed in Sections **5** and **6**.

Determining when a simple urn model such as the ones given above can be found to describe the extreme partially exchangeable probabilities may be difficult. For instance, we do not know how to extend the urn model for Markov chains to processes taking three values.

5.5 Technical Interpolation: Infinite Forms

Let us examine the results and problems for infinite sequences in the two preceding examples.

(10) Example—2 × 2 tables

Let $X_1, X_2, X_3, \ldots$ be an infinite sequence of random variables, each taking values 0 or 1. Suppose each trial is labeled as Male or Female and as Treated or Untreated and that the number of labels in each of the four possible categories (M, U), (M, T), (F, U), (F, T) is infinite. Suppose that for each n the distribution of $X_1, X_2, \ldots, X_n$ is partially exchangeable with respect to the suffcient statistic that counts the number of zeros and the number of ones for each label. Thus, $T_n = (a_1, b_1, a_2, b_2, a_3, b_3, a_4, b_4)$ where, for example, a_1 is the number of ones labeled (M, U), b_1 is the number of zeros labeled (M, U), a_2 is the number of ones labeled (M, T), b_2 is the number of zeros labeled (M, T), and so on. Then there is a unique probability distribution μ such that for every n and each sequence $x_1, x_2, \ldots, x_n$

(11) $pr(X_1 = x_1, \ldots, X_n = x_n) = \int \sum_{i=1}^{4} p_i^{a_i}(1-p_i)^{b_i}\, d\mu(p_1, p_2, p_3, p_4)$

where a_i, b_i are the values of $T_n(x_1, x_2, \ldots, x_n)$. The result can be proved by passing to the limit in the urn model (8) of **5.4**.[20]

(12) Example—Markov chains

Let $X_1, X_2, \ldots$ be an infinite sequence of random variables each taking values 0 or 1. For simplicity, assume $X_1 = 0$. Suppose that for each n the

[20] A different proof is given in G. Link's chapter in R. Jeffrey (ed.) (1980).

joint distribution of $X_1, \ldots, X_n$ is partially exchangeable with respect to the matrix of transition counts. Suppose that the following recurrence condition is satisfied

(13) $P(X_n = 0 \text{ infinitely often}) = 1.$

Then there is a unique probability distribution p such that for every n, and each sequence $x_2, x_3, \ldots, x_n$ of zeros and ones

(14) $P(X_1 = 0, X_2 = x_2, \ldots, X_n = x_n) = \int p_{11}^{t_{11}}(1 - p_{11})^{t_{10}} p_{00}^{t_{00}}(1-p_{00})^{t_{01}} d\mu(p_{11}, p_{00})$

where t_{ij} are the four entries of the transition matrix of $x_1, x_2, \ldots, x_n$.

De Finetti appears to state (pp. 218–219 of de Finetti [1974]) that the representation (14) is valid for every partially exchangeable probability assignment in this case. Here is an example to show that the representation (14) need not hold in the absence of the recurrence condition (13). Consider the probability assignment which goes 001111111... (all ones after two zeros) with probability one. This probability assignment is partially exchangeable and not representable in the form (14). It is partially exchangeable because, for any it, the first it symbols of this sequence form the only sequence with transition matrix $\left(\begin{smallmatrix} 1 & 1 \\ 0 & n-3 \end{smallmatrix}\right)$. To see that it is not representable in the form (14), write p_k for the probability that the last zero occurs at trial $k(k = 1, 2, 3, \ldots)$. For a mixture of Markov chains the numbers p_k can be represented as

(15) $p_k = c \int_0^1 p_{00}^k (1 - p_{00}) d\mu(p_{00}) \quad k = 1, 2, \ldots,$

where c is the probability mass the mixing distribution puts on $p_{11} = 1$. The representation (15) implies that the numbers p_k are decreasing. For the example 001111111..., $p_1 = 0, p_2 = 1, p_3 = p_4 = \ldots = 0$. So this example doesn't have a representation as a mixture of Markov chains. A detailed discussion of which partially exchangeable assignments are mixtures of Markov chains is in Diaconis and Freedman (1978b).

When can we hope for representations like (3), (11), and (14) in terms of averages over a naturally constructed "parameter space"?

The theory we have developed for such representations still uses the notation of a statistic T. Generally, the statistic will depend on the sample size n and one must specify the way the different T_n's interrelate.[21]

[21] Freedman (1962b) introduced the notion of S structures—in which, roughly,

Lauritzen (1973) compares S-structure with several other ways of linking together suffcient statistics. We have worked with a somewhat different notion in Diaconis and Freedman (1978c) and developed a theory general enough to include a wide variety of statistical models.

De Finetti sketches out what appears to be a general theory in terms of what he calls the *type* of an observation. In de Finetti (1938) he only gives examples which generalize the examples of 2×2 tables. Here things are simple. In the example there are four types of observations depending on the labels $(M, U), (M, T), (F, U), (F, T)$. In general, for each i we observe the value of another variable giving information like sex, time of day, and so on. An analog of (11) clearly holds in these cases. In de Finetti (1974), Section 9.6.2, the type of an observation is allowed to depend on the outcome of past observations as in our Markov chain example. In this example there are three types of observations—observations X_i that follow a 0 are of type zero; observations X_i that follow a 1 are of type one; and the first observation X_1 is of type 2. For the original case of exchangeability there is only one type.

In attempting to define a general notion of type we thought of trying to define a *type function* $t_n : 2^n \to \{1, 2, \ldots, r\}^n$ which assigns each symbol in a string a "type" depending only on past symbols. If there are r types, then it is natural to consider the statistic $T_n(x_1, \ldots, x_n) = (a_1, b_1, \ldots, a_r, b_r)$, where a_i, is the number of ones of type i and b_i is the number of zeros of type i. We suppose that the type functions t_n were chosen so that the statistics T_n had what Lauritzen (1973) has called $\sum$ structure—T_{n+1} can be computed from T_n and x_{n+1}. Under these circumstances we hoped that a probability *pr* which was partially exchangeable with respect to the T_n's would have a representation of the form

$$\textbf{(16)} \quad pr\{X_1 = x_1, \ldots, X_n = x_n\} = \int \prod_{i=1}^{n} p_i^{ai}(1 - p_i)^{b_i} d\mu(p_1, \ldots, p_r)$$

for some measure μ.

The following contrived counterexample indicates the difficulty. There will be two types of observations. However, there will be some partially exchangeable probabilities which fail to have the representation (16), even though the number of observations of each type becomes infinite.

EXAMPLE, You are observing a sequence of zeros and ones $X_0, X_1, X_2,$

from the value of $T_n(x_1, x_2, \ldots, x_n)$ and $T_m(x_{n+1}, x_{n+2}, \ldots, x_{n+m})$ one can compute $T_{n+m}(x_1, x_2, \ldots, x_{n+m})$.

$X_3, \ldots$ You know that one of 3 possible mechanisms generate the process:

- All the positions are exchangeable.
- The even positions $X_0, X_2, X_4, \ldots,$ are exchangeable with each other and the odd positions are generated by reading off a fixed reference sequence x of zeros and ones.
- The even positions $X_0, X_2, X_4, \ldots,$ are exchangeable with each other and the odd positions are generated by reading off the complement $\bar{x}$ (the complement has $\bar{x}_j = 1 - x_j$).

Knowing that the reference sequence is $x = 11101110\ldots$, *you keep track of two types of* X_i. If i is odd and all the preceding X_j with j odd lead to a sequence which matches x or $\bar{x}$, you call i of type 1. In particular, X_1 and X_3 are always of type 1. You count the number of zeros and ones of each type. Let $T_n = (a_1, b_1, a_2, b_2)$ be the type counts at time n. Any process of the kind described above is partially exchangeable for these T_n. Moreover, the sequence T_n has $\sum$ structure. Now consider the process which is fair coin tossing on the even trials and equals the ith coordinate of x on trial $2i - 1$. The number of zeros and ones of each type becomes infinite for this process. However, the process cannot be represented in the form (16). The class of all processes with these T_n's is studied in Diaconis and Freedman (1978c) where the extreme points are determined.

We do not know for which type functions the representation (16) will hold. Some cases when parametric representation is possible have been determined by Martin-Löf (1970, 1974) and Lauritzen (1976). A general theory of partial exchangeability and more examples are in Diaconis and Freedman (1978c).

5.6 Concluding Remarks

1. Some Bayesians are willing to talk about "tossing a coin with unknown p". For them, de Finetti's theorem can be interpreted as follows: If a sequence of events is exchangeable, then it is like the successive tosses of a p-coin with unknown p. Other Bayesians do not accept the idea of p coins with unknown p: de Finetti is a prime example. Writers on subjective probability have suggested that de Finetti's theorem bridges the gap between the two positions. We have trouble with this synthesis and the

following quote indicates that de Finetti has reservations about it:[22]

> The sensational effect of this concept (which went well beyond its intrinsic meaning) is described as follows in Kyburg and Smokler's preface to the collection *Studies in subjective probability* which they edited (pp. 13–14).
> In a certain sense the most important concept in the subjective theory is that of "exchangeable events". Until this was introduced (by de Finetti, 1931) the subjective theory of probability remained little more than a philosophical curiosity. None of those for whom the theory of probability was a matter of knowledge or application paid much attention to it. But, with the introduction of the concept of "equivalence or symmetry" or "exchangeability", as it is now called, a way was discovered to link the notion of subjective probability with the classical problem of statistical inference.

It does not seem to us that the theorem explains the idea of a coin with unknown p. The main point is this: Probability assignments involving mixtures of coin tossing were used by Bayesians long before de Finetti. The theorem gives a characterizing feature of such assignments—exchangeability—which can be thought about in terms involving only opinions about observable events.

The connection between mixtures and exchangeable probability assignments allows a subjectivist to interpret some of the classical calculations involving mixtures. For example, consider Laplace's famous calculation of the chance that the Sun will rise tomorrow given that it has risen on $n-1$ previou sdays (**5.1.3,** example 2). Laplace took a uniform prior on [0, 1] and calculated the chance as the posterior mean of the unknown parameter:

$$pr(\text{Sun rises tomorrow} \mid \text{Has risen on } n-1 \text{ previou sdays}) =$$

$$\frac{\int_0^1 p^n d\mu}{\int_0^1 p^{n-1} d\mu} = \frac{n}{n+1}.$$

The translation of this calculation is as follows: Let $X_i = 1$ if the Sun rises on day i, $X_i = 0$ otherwise. Laplace's uniform mixture of coin tossing is exchangeable and $P(X_1 = 1, \ldots, X_k =$

[22] 'Probability: Beware of falsifications!' *Scientia* **111** (1976) 283–303.

$1) = \frac{1}{k+1}, k = 1, 2, 3, \ldots$. With this allocation, $P(X_n = 1 | X_1 = 1, \ldots, X_{n-1} = 1) = \frac{P(X_1=1 \ldots X_n=1)}{P(X_1=1,\ldots,X_{n-1}=1)} = \frac{n}{n+1}$. A similar interpretation can be given to any calculation involving averages over a parameter space.

Technical Interpolations (2–4)

2. The exchangeable form of de Finetti's theorem (1) is also a useful computational device for specifying probability assignments. The result is more complicated and much less useful in the case of real valued variables. Here there is no natural notion of a parameter p. Instead, de Finetti's theorem says real valued exchangeable variables, $\{Xi\}_{i=1}^{\infty}$, are described as follows: There is a prior measure π on the space of all probability measures on the real line such that $pr(X_1 \in A_1, \ldots, X_n \in A_n) = \int \prod_{i=1}^{n} p(X_i \in A_i) d\pi(p)$. The space of all probability measures on the real line is so large that it is difficult for subjectivists to describe personally meaningful distributions on this space.[23] Thus, for real valued variables, de Finetti's theorem is far from an explanation of the type of parametric estimation—involving a family of probability distributions parametrized by a finite dimensional parameter space—that Bayesians from Bayes and Laplace to Lindley have been using in real statistical problems.

   In some cases, more restrictive conditions than exchangeability are reasonable to impose, and do single out tractable classes of distributions. Here are some examples adapted from Freedman.[24]

   Example: Scale mixtures of normal variables

   When can a sequence of real valued variables $\{X_i\} 1 \leq i < \infty$ be represented as a scale mixture of normal variables

$$\textbf{(19)} \quad pr(X_1, \leq t_1, \ldots, X_n \leq t_n) = \int_0^{\infty} \prod_{i=1}^{n} \Phi(\delta t_i) d\pi(\delta)$$

   where $\Phi(t) = \frac{1}{\sqrt{2\pi}} \int_{-\infty}^{t} e^{t^2/2} dt$?

   In Freedman (1963) it is shown that a necessary and suffcient condition for (19) to hold is that for each n the joint distribution

[23] Ferguson, "Prior distributions on spaces of probability measures", *Ann. Stat.* **2** (1974) 615–629 contains examples of the various attempts to choose such a prior.

[24] Freedman, "Mixtures of Markov processes", *Ann. Math. Stat.* **33** (1962a) 114–118, "Invariants under mixing which generalize de Finetti's theorem", *Ann. Math. Stat.* **33** (1962b) 916–923.

of $X_1, X_2, \ldots, X_n$ be rotationally symmetric. This result is related to the derivation of Maxwell's distribution for velocity in a Monatomic Ideal Gas.[25]

Example: Poisson distribution

Let $X_i (1 \leq i < \infty)$ take integer values. In Freedman (1962) it is shown that a necessary and suffcient condition for X_i to have a representation as a mixture of Poisson variables,

$$pr(X_1 = a_1, \ldots, X_n = a_n) = \int_0^\infty \prod_{i=1}^n e^{-\lambda} \frac{\lambda^{a_i}}{a_i!} d\pi(\lambda)$$

is as follows: For every n, the joint distribution of $X_1, X_2, \ldots, X_n$ given $S_n = \sum_{i=1}^n X_i$ must be multinomial, like the joint distribution of S_n balls dropped at random into n boxes.

Many further examples and some general theory are given in Diaconis and Freedman (1978c).

3. De Finetti's generalizations of partial exchangeability are closely connected to work of P. Martin-Löf's in the 1970's.[26] Martin-Löf's does not seem to work in a Bayesian context, rather he takes the notion of suffcient statistic as basic and from this constructs the joint distribution of the process in much the same way as de Finetti. Martin-Löf's connects the conditional distribution of the process with the microcanonical distributions of statistical mechanics. The idea is to specify the conditional distribution of the observations given the suffcient statistics. In this paper, and in Martin-Löf's work, the conditional distribution has been chosen as uniform. We depart from this assumption in Diaconis and Freedman (1978c) as we have in the example of mixtures of Poisson distributions. The families of conditional distributions we work with still "project" in the right way. This projection property allows us to use the well-developed machinery of Gibbs states as developed by Lanford (1973), Folmer (1974), and Preston (1977).

[25] Khinchin, *Mathematical foundations of statistical mechanics* (1949), chap. VI.

[26] Martin-Löf's work appears most clearly spelled out in a set of mimeographed lecture notes (Martin-Löf [1970]), unfortunately available only in Swedish. A technical treatment of part of this may be found in Martin-Löf (1974). Further discussion of Martin-Löf's ideas are in Lauritzen (1973, 1975) and the last third of Tjur (1974). These references contain many new examples of partially exchangeable processes and their extreme points. We have tried to connect Martin-Löf's treatment to the general version of de Finetti's theorem which we have derived in Diaconis and Freedman (1978c).

4. In this paper we have focused on generalizations of exchangeability involving a statistic. A more general extension involves the idea of invariance with respect to a collection of transformations of the sample space into itself. This contains the idea of partial exchangeability with respect to a statistic since we can consider the class of all transformations leaving the statistic invariant. A typical case which cannot be neatly handled by a finite dimensional statistic is de Finetti's theorem for zero/one matrices. Here the distribution of the doubly infinite random matrix is to be invariant under permutations of the rows and columns. David Aldous has recently identified the extreme points of these matrices and shown that no representation as mixture of finite parameter processes is possible.
5. Is exchangeability a natural requirement on subjective probability assignments? It seems to us that much of its appeal comes from the (forbidden?) connection with coin tossing. This is most strikingly brought out in the thumbtack Markov chain example. If someone were thinking about assigning probability to 10 flips of the tack and had never heard of Markov chains it seems unlikely that they would hit on the appropriate notion of partial exchangeability. The notion of symmetry seems strange at first. Its appeal comes from the connection of Markov chains with unknown transition probabilities. A feeling of naturalness only appears after experience and refection.

6

Choosing

This is an exposition of the framework for decision-making that Ethan Bolker and I floated about 40 years ago, in which options are represented by propositions and any choice is a decision to make some proposition true.[1] It now seems to me that the framework is pretty satisfactory, and I shall present it here on its merits, without spending much time defending it against what I have come to see as fallacious counterarguments.

6.1 Preference Logic

To the option of making the proposition A true corresponds the conditional probability distribution $pr(\cdots|A)$, where the unconditional distribution $pr(\cdots)$ represents your prior probability judgment—prior, that is, to deciding which option-proposition to make true. And your expectation of utility associated with the A-option will be your conditional expectation $ex(u|A)$ of the random variable u (for "utility"). This conditional expectation is also known as your **desirability** for truth of A, and denoted '$desA$':

$$desA = ex(u|A)$$

[1] See Bolker (1965, 1966, 1967) and Jeffrey (1965, 83, 90, 96). When Bolker and I met, in 1963, I had worked out the logic as in **6.1** and was struggling to prove what turned out to be the wrong uniqueness theorem, while Bolker proved to have found what I needed: a statement and proof of the right uniqueness theorem (along with a statement and proof of the corresponding *existence* theorem—see chapter 9 of Jeffrey 1983 or 1990).

Now preference ($\succ$), indifference ($\approx$), and preference-or-indifference ($\succeq$) go by desirability, so that

$$A \succ B \quad \text{if} \quad desA > desB,$$
$$A \approx B \quad \text{if} \quad desA = desB,$$
$$A \succsim B \quad \text{if} \quad desA \geq desB,$$

and similarly for $\prec$ and $\precsim$. Note that it is not only option-propositions that appear in preference rankings; you can perfectly well prefer a sunny day tomorrow (truth of 'Tomorrow will be sunny') to a rainy one even though you know you cannot affect the weather.[2]

The basic connection between *pr* and *des* is that their product is additive, just as *pr* is.[3] Here is another way of saying that, in a different notation:

Basic Probability-Desirability Link

If the S_i form a partitioning of $\top$, then

(1) $des(A) = \sum_i des(S_i \wedge A)pr(S_i|A)$.

Various principles of preference logic can now be enunciated, and fallacies identified, as in the following samples. The first is a fallacious mode of inference according to which denial reverses preferences. (The symbol '$\vdash$' means valid implication: that the premise, to the left of the turnstyle, validly implies the conclusion, to its right.)

6.1.1 *Denial Reverses Preferences:* $A \succ B \vdash \neg B \succ \neg A$

COUNTEREXAMPLE: **Death before dishonor.** Suppose you are an exemplary Roman of the old school, so that if you were dishonored you would certainly be dead—if necessary, by suicide: $pr(\text{death}|\text{dishonor}) = 1$. Premise: $A \succ B$, you prefer death (A) to dishonor (B). Then **1.1.1** is just backward; it must be that you prefer a guarantee of life ($\neg A$) to a guarantee of not being dishonored ($\neg B$), for in this particular example your conditional probability $pr(A|B) = 1$ makes the second guarantee an automatic consequence of the first.

[2] Rabinowicz (2002) effectively counters arguments by Levi and Spohn that Dutch Book arguments are untrustworthy in cases where the bet is on the agent's own future betting behavior, since "practical deliberation crowds out self-prediction" much as the gabble of cell-phone conversations may make it impossible to pursue simple trains of thought in railway cars. The difficulty is one that cannot arise where it is this very future behavior that is at issue.

[3] Define $\tau(H) = pr(H)des(H)$; then $\tau(H_1 \vee H_2 \vee \ldots) = \sum_i \tau(H_i)$ provided $\tau(H_i \wedge H_j) = 0$ whenever $i \neq j$.

6.1.2 If $A \wedge B \vdash \bot$ *and* $A \succeq B$ *then* $A \succeq A \vee B \succeq B$

Proof. Set $w = pr(A|A \vee B)$. Then $des(A \vee B) = w(desA) + (1 - w)(desB)$. This is a convex combination of $desA$ and $desB$, which must therefore lie either between them or at one or the other. This one is valid.

6.1.3 The "Sure Thing" (or "Dominance") Principle: $(A \wedge B) \succ C, (A \wedge \neg B) \succ C \vdash A \succ C$

In words, with premises to the left of the turnstyle: If you would prefer A to C knowing that B is true, and knowing that B is false, you prefer A to C.

This snappy statement is not a valid form of inference. *Really?* If A is better than C no matter how other things (B) turn out, then surely A is as good as B. How could that be wrong?

Well, here is one way, with $C =$ Quit smoking in **6.1.3:**

COUNTEREXAMPLE, **The Marlboro Man, "MM".** He reasons: 'Sure, smoking (A) makes me more likely to die before my time (B), but a smoker can live to a ripe old age. Now if I am to live to a ripe old age I'd rather do it as a smoker, and if not I would still enjoy the consolation of smoking. *In either case*, I'd rather smoke than quit. So I'll smoke.' In words, again:

> PREMISE: Smoking is preferable to quitting if I die early.
> PREMISE: Smoking is preferable to quitting if I do not die early.
> CONCLUSION: Smoking is preferable to quitting.

I have heard that argument put forth quite seriously. And there are versions of decision theory which seem to endorse it, but the wise heads say 'NONSENSE, DEAR BOY,' you have chosen a partition $\{B, \neg B\}$ relative to which the STP fails. But there are provisos under which the principle is quite trustworthy. Sir Ronald Fisher (1959) actually sketched a genetic scenario in which that would be the case. (See **6.3.1** below.) But he never suggested that **6.1.3**, the unrestricted STP, is an a priori justification for the MM's decision.

For the record, this is the example that L. J. Savage (1954), the man who brought us the STP, used to motivate it.

> A businessman contemplates buying a certain piece of property. He considers the outcome of the next presidential election relevant to the attractiveness of the purchase. So, to clarify the matter for himself, he asks whether he would buy if he knew that the Republican candidate were going to win, and decides that he would do so. Similarly, he considers whether he would buy if he knew that the Democratic candidate were going to win, and again finds that he would do so. Seeing that he would buy in either event, he decides that he should buy, even though he does not know which event obtains, or will obtain, as we would ordinarily say. It is all too seldom that a decision can be arrived at on the basis of the principle used by this businessman, but, except possibly for the assumption of simple ordering, I know of no other extralogical principle governing decisions that finds such ready acceptance. (Savage 1954, p. 21)

Savage's businessman and my Marlboro Man seem to be using exactly the same extralogical principle (deductive logic, that is). What is going on?

6.1.3.1 BJ-ing the MM

This is how it normally turns out that people with the MM's preferences prefer quitting to smoking—even if they cannot bring themselves to make the preferential choice. It's all done with conditional probabilities. Grant the MM's preferences and probabilities as on the following assumptions about desirabilities and probabilities, where 'long' and 'shorter' refer to expected length of life, and w, s, l ("worst, smoke, long") are additive components of desirability where $s, l > 0$ and $w + s + l < 1$.

$$\begin{aligned} des(long \wedge smoke) &= l + s + w & pr(long|smoke) &= p \\ des(long \wedge quit) &= l + w & pr(long|quit) &= q \\ des(shorter \wedge smoke) &= s + w & pr(shorter|smoke) &= 1 - p \\ des(shorter \wedge quit) &= w & pr(shorter|quit) &= 1 - q \end{aligned}$$

Question: How must MM's conditional probabilities p, q be related in order for the B-J figure of merit to advise his quitting instead of continuing to smoke?

Answer: $des(quit) > des(smoke)$ if and only if $(l + s + w)p + (s + w)$ $(1 - p) > (l + w)q + w(1 - q)$. Then BJ tells MM to smoke iff $p - q < \frac{s}{l}$.

These are the sorts of values of p and q that MM might have gathered from Consumers Union (1963): $p = .59$, $q = .27$. Then $.32 > s/l$ where, by hypothesis, s/l is positive and less than 1; and as long as MM is not so addicted or otherwise dedicated to smoking that the increment of desirability from it is at least 32% of the increment from long life, formula (1) in the basic probability-desirability link will have him choose to quit.

6.1.3.2 BJ-ing the STP

If we interpret Savage (1954) as accepting the basic probability-desirability link (1) as well as the STP, he must be assuming that acts are probabilistically independent of states of nature. The businessman must then consider that his buying or not will have no influence on the outcome of the election; he must then see p and q as equal, and accept the same reasoning for the MM and the businessman: choose the dominant act (smoke, buy).[4]

6.1.4 Bayesian Frames

Given a figure of merit for acts, Bayesian decision theory represents a certain structural concept of rationality. This is contrasted with substantive criteria of rationality having to do with the aptness of particular probability and utility functions in particular predicaments. With Donald Davidson[5] I would interpret this talk of rationality as follows. What remains when all substantive questions of rationality are set aside is bare logic, a framework for tracking your changing judgments, in which questions of validity and invalidity of argument-forms can be settled as illustrated above. The discussion goes one way or another, depending on particulars of the figure of merit that is adopted for acts. Here we compare and contrast the Bolker-Jeffrey figure of merit with Savage's.

A complete, consistent set of substantive judgments would be represented by a "Bayesian frame" consisting of (1) a probability distribution over a space Ω of "possible worlds", (2) a function u assigning "utilities" $u(\omega)$ to the various worlds in $\omega \in \Omega$, and (3) an assignment of subsets of Ω as values of the sentence-letters 'A', 'B',.... Then subsets $A \subseteq \Omega$ represent propositions; A is true in world ω iff $\omega \in A$; and conjunction, disjunction, and denial of propositions is represented by the intersection, union, and complement with respect to Ω of the corresponding subsets.

[4] This is a reconstructionist reading of Savage (1954), not widely shared.
[5] Davidson (1980) 273–274. See also Jeffrey (1987) and sec. 12.8 of Jeffrey (1965, 1983, 1990).

In any logic, validity of an argument is truth of the conclusion in every frame in which all the premises are true. In a Bayesian logic of decision, Bayesian frames represent possible answers to substantive questions of rationality; we can understand that, without knowing how to determine whether particular frames would be substantively rational for you on particular occasions. So in Bayesian decision theory we can understand validity of an argument as truth of its conclusion in any Bayesian frame in which all of its premises are true, and understand consistency of a judgment (say, affirming $A \succ B$ while denying $\neg B \succ \neg A$) as existence of a non-empty set of Bayesian frames in which the judgment is true. On this view, bare structural rationality is simply representability in the Bayesian framework.

6.2 Causality

In decision-making it is deliberation, not observation, that changes your probabilities. To think you face a decision problem rather than a question of fact about the rest of nature is to expect whatever changes arise in your probabilities for those states of nature during your deliberation to stem from changes in your probabilities of choosing options. In terms of the analogy with mechanical kinematics: As a decision-maker you regard probabilities of options as inputs driving the mechanism, not driven by it.

Is there something about your judgmental probabilities which shows that you are treating truth of one proposition as promoting truth of another—rather than as promoted by it or by truth of some third proposition which also promotes truth of the other? Here the promised positive answer to this question is used to analyze puzzling problems in which we see acts as mere symptoms of conditions we would promote or prevent if we could. Such "Newcomb problems" (Nozick, 1963, 1969, 1990) seem to pose a challenge to the decision theory floated in Jeffrey (1965, 1983, 1990), where notions of causal influence play no rôle. The present suggestion about causal judgments will be used to question the credentials of Newcomb problems as decision problems.

The suggestion is that imputations of causal influence do not show up simply as momentary features of probabilistic states of mind, but as intended or expected features of their evolution. Recall the following widely recognized necessary condition for the judgment that truth of one proposition ("cause") promotes truth of another ("effect"):

Correlation: $pr(\text{effect}|\text{cause}) > pr(\text{effect}|\neg\text{cause})$.

But aside from the labels, what distinguishes cause from effect in this relationship? The problem is that it is a symmetrical relationship, which continues to hold when the labels are interchanged: $pr(\text{cause}|\text{effect}) > pr(\text{cause}|\neg\text{effect})$. It is also problematical that correlation is a relationship between contemporaneous values of the same conditional probability function.

What further condition can be added, to produce a necessary and sufficient pair? With Arntzenius (1990), I suggest the following answer, i.e., rigidity relative to the partition $\{\text{cause}, \neg\text{cause}\}$.[6]

Rigidity. Constancy of $pr(\text{effect}|\text{cause})$
and $pr(\text{effect}|\neg\text{cause})$ as $pr(\text{cause})$ varies.

Rigidity is a condition on a variable ('*pr*') that ranges over a set of probability functions. The functions in the set represent ideally definite momentary probabilistic states of mind for the deliberating agent, as they might be at different times. This is invariance of conditional probabilities (**3.1**), shown in **3.2** to be equivalent to probability kinematics as a mode of updating. In statistics, the corresponding term is 'sufficiency': A sufficient statistic is a random variable whose sets of constancy ("data") form a partition satisfying the rigidity condition. Clearly, *pr* can vary during deliberation, for if deliberation converges toward choice of a particular act, the probability of the corresponding proposition will rise toward 1. In general, agents' intentions or assumptions about the kinematics of *pr* might be described by maps of possible courses of evolution of probabilistic states of mind—often, very simple maps. These are like road maps in that paths from point to point indicate feasibility of passage via the anticipated mode of transportation, e.g., ordinary automobiles, not "all terrain" vehicles. Your kinematical map represents your understanding of the dynamics of your current predicament, the possible courses of development of your probability and desirability functions.

The Logic of Decision used conditional expectation of utility given an act as the figure of merit for the act, namely, its desirability, *des*(act). Newcomb problems (Nozick 1969) led many to see that figure as acceptable only on special causal assumptions, and a number of versions of

[6] In general the partition need not be twofold. Note that if 'cause' denotes one element of a partition and '¬cause' denotes the disjunction of all other elements, ¬cause need not satisfy the rigidity condition even though all elements of the original partition do.

"causal decision theory" have been proposed as more generally acceptable.[7]

(CFM) Causal figure of merit relative to $\mathcal{K} = ex_{\mathcal{K}}(u|A)$.

In the discrete case, $\mathcal{K} = \{K_k : k = 1, 2, \ldots\}$ and the CFM is $\sum_k des(A \wedge K_k)$. But if Newcomb problems are excluded as bogus, then in genuine decision problems *des* (act) will remain constant throughout deliberation, and will be an adequate figure of merit.

Now there is much to be said about Newcomb problems, and I have said quite a lot elsewhere, as in Jeffrey (1996), and I am reluctant to finish this book with close attention to what I see as a side-issue. I would rather finish with an account of game theory, what Robert Aumann calls "interactive decision theory". But I don't seem to have time for that, and if you have not been round the track with Newcomb problems you may find the rest worthwhile, so I leave much of it in, under the label "supplements".

6.3 Supplements: Newcomb Problems

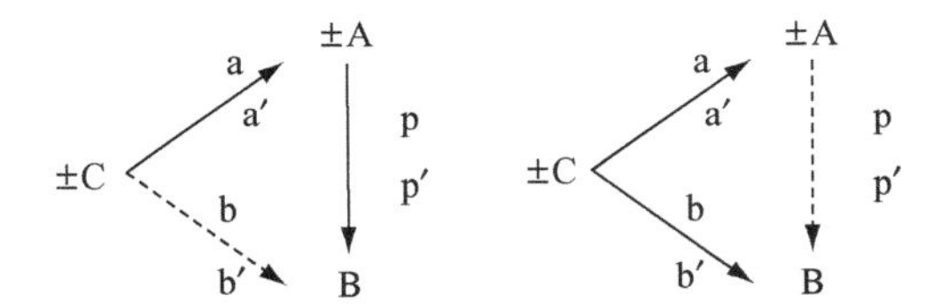

(a) Ordinarily, acts ±A screen off deep states ±C from plain states ±B.

(b) In Newcomb problems deep states screen off acts from plain states.

Figure 1. Solid/dashed arrows indicate stable/labile conditional probabilities.

PREVIEW. The simplest sort of decision problem is depicted in Fig. 1(a), where C and $\neg C$ represent states of affairs that may incline you one way or the other between options A and $\neg A$, and where your choice between those options may tend to promote or prevent the state of affairs B. But there is no probabilistic causal influence of $\pm C$ directly on $\pm B$, without passing through $\pm A$. Fig. 1(b) depicts the simplest "Newcomb" problem, where the direct probabilistic influence runs from $\pm C$ directly

[7] In the one I like best (Skyrms 1980), the figure of merit for choice of an act is the agent's expectation of $des(A)$ on a partition $\mathcal{K}$ of causal hypotheses.

to $\pm A$ and to $\pm B$, but there is no direct influence of $\pm A$ on $\pm B$. Thus, Newcomb problems are not decision problems.

> For you to view $\pm A$ as a direct probabilistic influence on truth or falsity of B, your conditional probabilities $p = pr(B|A)$ and $p' = pr(B|\neg A)$ need to remain stable as your conditional probability $pr(A)$ varies.

6.3.1 "The Mild and Soothing Weed"

For smokers who see quitting as prophylaxis against cancer, preferability goes by initial *des*(act) as in Fig 1a; but there are views about smoking and cancer on which these preferences might be reversed. Thus, R. A. Fisher (1959) urged serious consideration of the hypothesis of a common inclining cause of (A) smoking and (B) bronchial cancer in (C) a bad allele of a certain gene, possessors of which have a higher chance of being smokers and of developing cancer than do possessors of the good allele (independently, given their allele). On that hypothesis, smoking is bad news for smokers but not bad for their health, being a mere sign of the bad allele, and so, of bad health. Nor would quitting conduce to health, although it would testify to the agent's membership in the low-risk group.

On Fisher's hypothesis, where acts $\pm A$ and states $\pm B$ are seen as independently promoted by genetic states $\pm C$, i.e., by presence (C) or absence ($\neg C$) of the bad allele of a certain gene, the kinematical constraints on pr are the following. (Thanks to Brian Skyrms for this.)
Rigidity: The following are constant as $c = pr(C)$ varies.

$$a = pr(A|C), \quad a' = pr(A|\neg C), \quad b = pr(B|C), \quad b' = pr(B|\neg C)$$

Correlation:

$$p > p', \text{ where } p = pr(B|A) \text{ and } p' = pr(B|\neg A)$$

Indeterminacy: None of a, b, a', b' are 0 or 1,
Independence: $pr(A \wedge B|C) = ab$, $pr(A \wedge B|\neg C) = a'b'$.

Since in general, $pr(F|G \wedge H) = pr(F \wedge G|H)/pr(G|H)$, the independence and rigidity conditions imply that $\pm C$ screens off A and B from each other in the following sense.
Screening-off:

$$pr(A|B \wedge C) = a, \quad pr(A|B \wedge \neg C) = a',$$
$$pr(B|A \wedge C) = b, \quad pr(B|A \wedge \neg C) = b'.$$

Under these constraints, preference between A and $\neg A$ can change as $pr(C) = c$ moves out to either end of the unit interval in thought-experiments addressing the question "What would $des(A) - des(\neg A)$ be if I found I had the bad/good allele?" To carry out these experiments, note that we can write $p = pr(B|A) = pr(A \wedge B)/pr(A) = \frac{pr(A|B \wedge C)pr(B|C)pr(C) + pr(A|B \wedge \neg C)pr(B|\neg C)pr(\neg C)(1-c)}{pr(A|C)pr(C) + pr(A|\neg C)pr(\neg C)}$ and similarly for $p' = pr(B|\neg A)$. Then we have

$$p = \frac{abc + a'b'(1-c)}{ac + a'(1-c)}, \qquad p' = \frac{(1-a)bc + (1-a')b'(1-c)}{(1-a)c + (1-a')(1-c)}.$$

Now final p and p' are equal to each other, and to b or b' depending on whether final c is 1 or 0. Since it is c's rise to 1 or fall to 0 that makes $pr(A)$ rise or fall as much as it can without going off the kinematical map, the (quasi-decision) problem has two ideal solutions, i.e., mixed acts in which the final unconditional probability of A is the rigid conditional probability, a or a', depending on whether c is 1 or 0. But $p = p'$ in either case, so each solution satisfies the conditions under which the dominant pure outcome (A) of the mixed act maximizes $des(\pm A)$. (This is a quasi-decision problem because what is imagined as moving c is not the decision but factual information about C.)

As a smoker who believes Fisher's hypothesis you are not so much trying to make your mind up as trying to discover how it is already made up. But this may be equally true in ordinary deliberation, where your question "What do I really want to do?" is often understood as a question about the sort of person you are, a question of which option you are already committed to, unknowingly. The diagnostic mark of Newcomb problems is a strange linkage of this question with the question of which state of nature is actual—strange, because where in ordinary deliberation any linkage is due to an influence of acts $\pm A$ on states $\pm B$, in Newcomb problems the linkage is due to an influence, from behind the scenes, of deep states $\pm C$ on acts $\pm A$ and plain states $\pm B$. This difference explains why deep states ("the sort of person I am") can be ignored in ordinary decision problems, where the direct effect of such states is wholly on acts, which mediate any further effect on plain states. But in Newcomb problems deep states must be considered explicitly, for they directly affect plain states as well as acts (Fig. 1).

In the kinematics of decision the dynamical role of forces can be played by acts or deep states, depending on which of these is thought to influence plain states directly. Ordinary decision problems are modeled kinematically by applying the rigidity condition to acts as causes. Ordinarily, acts screen off deep states from plain ones in the sense that B is conditionally independent of $\pm C$ given $\pm A$, so that while it is variation in c that makes $pr(A)$ and $pr(B)$ vary, the whole of the latter variation is accounted for by the former (Fig. 1a). But to model Newcomb problems kinematically we apply the rigidity condition to the deep states, which screen off acts from plain states (Fig. 1b). In Fig. 1a, the probabilities b and b' vary with c in ways determined by the stable a's and p's, while in Fig. 1b the stable a's and b's shape the labile p's as we have seen above:

$$p = \frac{abc + a'b'(1-c)}{ac + a'(1-c)}, \qquad p' = \frac{(1-a)bc + (1-a')b'(1-c)}{(1-a)c + (1-a')(1-c)}.$$

Similarly, in Fig. 1a the labile probabilities are

$$b = \frac{apc + a'p'(1-c)}{ac + a'(1-c)}, \qquad b' = \frac{(1-a)pc + (1-a')p'(1-c)}{(1-a)c + (1-a')(1-c)}.$$

While C and $\neg C$ function as causal hypotheses, they do not announce themselves as such, even if we identify them by the causal roles they are meant to play, as when we identify the "bad" allele as the one that promotes cancer and inhibits quitting. If there is such an allele, it is a still unidentified feature of human DNA. Fisher was talking about hypotheses that further research might specify, hypotheses he could only characterize in causal and probabilistic terms—terms like "malaria vector" as used before 1898, when the anopheles mosquito was shown to be the organism playing that aetiological rôle. But if Fisher's science fiction story had been verified, the status of certain biochemical hypotheses C and $\neg C$ as the agent's causal hypotheses would have been shown by satisfaction of the rigidity conditions, i.e., constancy of $pr(-|C)$ and of $pr(-|\neg C)$, with C and $\neg C$ spelled out as technical specifications of alternative features of the agent's DNA. Probabilistic features of those biochemical hypotheses, e.g., that they screen acts off from states, would not be stated in those hypotheses, but would be shown by interactions of those hypotheses with pr, B, and A, i.e., by truth of the following consequences of the kinematical constraints.

$$pr(B|\text{act} \wedge C) = pr(B|C), \quad P(B|\text{act} \wedge \neg C) = pr(B|\neg C).$$

No purpose would be served by packing such announcements into the hypotheses themselves, for even if true, such announcements would be redundant. The causal talk, however useful as commentary, does no work in the matter commented upon.[8]

6.3.2 The Flagship Newcomb Problem

Nozick's (1969) flagship Newcomb problem resolutely fends off naturalism about deep states, making a mystery of the common inclining cause of acts and plain states while suggesting that the mystery could be cleared up in various ways, pointless to elaborate. Thus, Nozick (1969) begins:

> Suppose a being [call her 'Alice'] in whose power to predict your choices you have enormous confidence. (One might tell a science-fiction story about a being from another planet, with an advanced technology and science, who you know to be friendly, and so on.) You know that this being has often correctly predicted your choices in the past (and has never, so far as you know, made an incorrect prediction about your choices), and furthermore you know that this being has often correctly predicted the choices of other people, many of whom are similar to you, in the particular situation to be described below. One might tell a longer story, but all this leads you to believe that almost certainly this being's prediction about your choice in the situation to be discussed will be correct. There are two boxes...

Alice has surely put \$1,000 in one box, and (B) she has left the second box empty or she has ($\neg B$) put \$1,000,000 in it, depending on whether she predicts that you will (A_2) take both boxes, or that youw ill (A_1) take only one—the one with \$1000 in it. (In the previous notation, $A_2 = A$.)

Here you are to imagine yourself in a probabilistic frame of mind where your desirability for A_1 is greater than that for A_2 because although you think A's truth or falsity has no influence on B's, your conditional probability $p = pr(B|A_1)$ is ever so close to 1 and your $p' = pr(B|A_2)$ is ever so close to 0. Does that seem a tall order? Not to worry! High correlation is a red herring; a tiny bit will do, e.g., if desirabilities are proportional to dollar payoffs, then the 1-box option, $\neg A$, maximizes desirability as long as $pr(B|A_1)$ is greater than .001.

[8] See Joyce (2002); also Leeds (1984) in another connection.

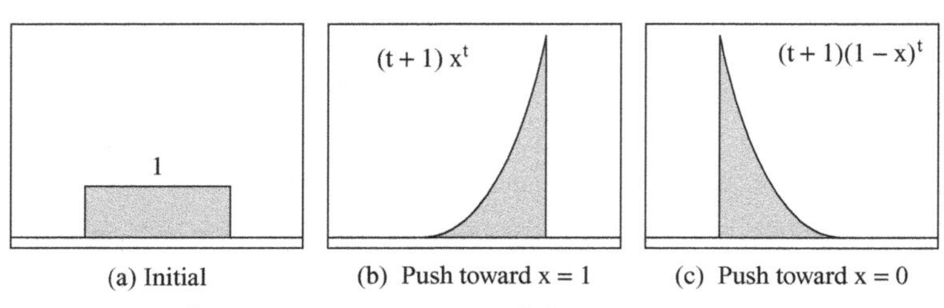

(a) Initial (b) Push toward x = 1 (c) Push toward x = 0

Figure 2. Two patterns of density migration

To see how that might go numerically, think of the choice and the prediction as determined by independent drawings by the agent and the predictor from the same urn, which contains tickets marked '2' and '1' in an unknown proportion x : 1-x. Initially, the agent's unit of probability density over the range [0,1] of possible values of x is flat (Fig. 2a), but in time it can push toward one end of the unit interval or the other, e.g., as in Fig. 2b, c.[9] At t = 997 these densities determine the probabilities and desirabilities in Table 3b and c, and higher values of t will make $desA_1 - desA_2$ positive. Then if t is calibrated in thousandths of a minute this map has the agent preferring the 2-box option after a minute's deliberation. The urn model leaves the deep state mysterious, but clearly specifies its mysterious impact on acts and plain states.

bad	best
worst	good

(a)

bad	best
worst	good

(b)

bad	best
worst	good

(c)

Table 3. Deterministic 1-box solution: $pr(\text{shaded}) = 0$;
beta = 1 until decision.

The irrelevant detail of high correlation or high $K = pr(B|A) - pr(B|\neg A)$, was a bogus shortcut to the 1-box conclusion, obtained if K is not just high but maximum, which happens when $p = 1$ and $p' = 0$. This means that the "best" and "worst" cells in the payoff table have unconditional probability 0. Then taking both boxes means a thousand, taking just one means a million, and preference between acts is clear, as long as $p - p'$ is neither 0 nor 1, and remains maximum, 1. The density functions of Fig. 2 are replaced by probability assignments r and $1 - r$ to the possibilities that the ratio of 2-box tickets to 1-box

[9] In this kinematical map $pr(A) = \int_0^1 x^{t+1} f(x)\, d(x)$ and $pr(B|A) = \int_0^1 x^{t+2} f(x)\, dx / pr(A)$ with $f(x)$ as in Fig. 2(b) or (c). Thus, with $f(x)$ as in (b), $pr(A) = (t+1)/(t+3)$ and $pr(B|A) = (t+2)/(t+3)$. See Jeffrey (1988).

density functions of Fig. 2 are replaced by probability assignments r and $1 - r$ to the possibilities that the ratio of 2-box tickets to 1-box tickets in the urn is 1:0 and 0:1, i.e., to the two ways in which the urn can control the choice and the prediction deterministically and in the same way. In place of the smooth density spreads in Fig. 2 we now have point-masses r and $1 - r$ at the two ends of the unit interval, with desirabilities of the two acts constant as long as r is neither 0 nor 1. Now the 1-box option is preferable throughout deliberation, up to the very moment of decision.[10] But of course this reasoning uses the premise that $pr(\text{predict 2}|\text{take 2}) - pr(\text{predict 2}|\text{take 1}) = \beta = 1$ through deliberation, a premise making abstract sense in terms of uniformly stocked urns, but very hard to swallow as a real possibility.

6.3.3 Hofstadter

Hofstadter (1983) saw prisoners' dilemmas as down-to-earth Newcomb problems. Call the prisoners Alma and Boris. If one confesses and the other does not, the confessor goes free and the other serves a long prison term. If neither confesses, both serve short terms. If both confess, both serve intermediate terms. From Alma's point of view, Boris's possible actions (B, confess, or $\neg B$, don't) are states of nature. She thinks they think alike, so that her choices (A, confess, $\neg A$, don't) are pretty good predictors of his, even though neither's choices influence the other's. If both care only to minimize their own prison terms this problem fits the format of Table 3(a). The prisoners are thought to share a characteristic determining their separate probabilities of confessing in the same way—independently, on each hypothesis about that characteristic. Hofstadter takes that characteristic to be rationality, and compares the prisoners' dilemma to the problem Alma and Boris might have faced as bright children, independently working the same arithmetic problem, whose knowledge of each other's competence and ambition gives them good reason to expect their answers to agree before either knows the answer: "If reasoning guides me to [...], then, since I am no different from anyone else as far as rational thinking is concerned, it will guide everyone to [...]." The deep states seem less mysterious here than in the flagship Newcomb problem; here they have some such form as Cx'' = We are both likely to get the right answer, i.e., x. (And here ratios of utilities

[10] At the moment of decision the desirabilities of shaded rows in (b) and (c) are not determined by ratios of unconditional probabilities, but continuity considerations suggest that they remain good and bad, respectively.

are generally taken to be on the order of 10:1 instead of the 1000:1 ratios that made the other endgame so demanding. With utilities $0, 1, 10, 11$ instead of $0, 1, 1000, 1001$, indifference between confessing and remaining silent now comes at =10% instead of one-tenth of 1%.) But to heighten similarity to the prisoners' dilemma let us suppose the required answer is the parity of x, so that the deep states are simply C = We are both likely to get the right answer, i.e., even, and $\neg C$ = We are both likely to get the right answer, i.e., odd.

What's wrong with Hofstadter's view of this as justifying the cooperative solution? [And with von Neumann and Morgenstern's (p. 148) transcendental argument, remarked upon by Skyrms (1990, 13–14), for expecting rational players to reach a Nash equilibrium?] The answer is failure of the rigidity conditions for acts, i.e., variability of pr(He gets x | I get x) with pr(I get x) in the decision maker's kinematical map. It is Alma's conditional probability functions $pr(-|\pm C)$ rather than $pr(-|\pm A)$ that remain constant as her probabilities for the conditions vary. The implausibility of initial *des*(act) as a figure of merit for her act is simply the implausibility of positing constancy of her probabilities conditional on her acts as her probability function pr evolves in response to changes in $pr(A)$. But the point is not that confessing is the preferable act, as causal decision theory would have it. It is rather that Alma's problem is not indecision about which act to choose, but ignorance of which allele is moving her.

Hofstadter's (1983) version of the prisoners' dilemma and the flagship Newcomb problem have been analyzed here as cases where plausibility demands a continuum [0,1] of possible deep states, with opinion evolving as smooth movements of probability density toward one end or the other draw probabilities of possible acts along toward 1 or 0. The problem of the smoker who believes Fisher's hypothesis was simpler in that only two possibilities $(C, \neg C)$ were allowed for the deep state, neither of which determined the probability of either act as 0 or 1.

6.3.4 Conclusion

The flagship Newcomb problem owes its bizarrerie to the straightforward character of the pure acts: Surely you can reach out and take both boxes, or just the opaque box, as you choose. Then as the pure acts are options, you cannot be committed to either of the non-optional mixed acts. But in the Fisher problem, those of us who have repeatedly "quit" easily appreciate the smoker's dilemma as humdrum entrapment in some mixed act, willy-nilly. That the details of the entrapment are describable as cycles of temptation, resolution and betrayal makes the history no less believable—only more petty. Quitting and continuing are not options, i.e., $prA \approx 0$ and $prA \approx 1$ are not destinations you think you can choose, given your present position on your kinematical map, although you may eventually find yourself at one of them. The reason is your conviction that if you knew your genotype, your value of $pr(A)$ would be either a or a', neither of which is ≈ 0 or ≈ 1. (Translation: "At places on the map where $pr(C)$ is at or near 0 or 1, prA is not.") The extreme version of the story, with $a \approx 1$ and $a' \approx 0$, is more like the flagship Newcomb problem; here you do see yourself as already committed to one of the pure acts, and when you learn which that is, you will know your genotype.

I have argued that Newcomb problems are like Escher's famous staircase on which an unbroken ascent takes you back where you started.[11] We know there can be no such things, but see no local flaw; each step makes sense, but there is no way to make sense of the whole picture; that's the art of it.[12]

In writing *The Logic of Decision* in the early 1960's I failed to see Bob Nozick's deployment of Newcomb's problem in his Ph.D. dissertation as a serious problem for the theory floated in chapter 1 and failed to see the material on rigidity in chapter 11 as solving that problem. It took a long time for me to take the "causal" decision theorists (Stalnaker, Gibbard and Harper, Skyrms, Lewis, Sobel) as seriously as I needed to in order to appreciate that my program of strict exclusion of causal talk would be an obstacle for me—an obstacle that can be overcome by recognizing causality as a concept that emerges naturally when we patch together the differently dated or clocked probability assignments that arise in formulating and then solving a decision problem. I now conclude that in Newcomb problems, "One box or two?" is not a question about how to choose, but about what you are already set to do, willy-nilly. Newcomb problems are not decision problems.

[11] Escher (1960), based on Penrose and Penrose (1958), 31–33; see Escher (1989), p. 78.

[12] Elsewhere I have accepted Newcomb problems as decision problems, and accepted "2-box solutions" as correct. In Jeffrey 1983, sec. 1.7 and 1.8, I proposed a new criterion of acceptability of an act—"ratifiability"—which proved to break down in certain cases (see the corrected paperback 2nd edition, 1990, p. 20). In Jeffrey (1988, 1993) ratifiability was recast in terms more like the present ones, but still treating Newcomb problems as decision problems—a position which I now renounce.

References

Frank Arntzenius, 'Physics and common causes', *Synthese* **82** (1990) 77–96.

Kenneth J. Arrow et al. (eds.), *Rational Foundations of Economic Behaviour*, New York, St. Martin's (1996).

Thomas Bayes, 'Essay toward solving a problem in the doctrine of chances,' *Philosophical Transactions of the Royal Society* **50** (1763), reprinted in *Facsimiles of Two Papers by Bayes*, New York, Hafner (1963).

Ethan Bolker, *Functions Resembling Quotients of Measures*, Ph.D. dissertation, Harvard University (1965).

_____, 'Functions resembling quotients of measures', *Transactions of the American Mathematical Society* **124** (1966) 292–312.

_____, 'A simultaneous axiomatization of utility and subjective probability', *Philosophy of Science* **34** (1967) 333–340.

A. Bruno, 'On the notion of partial exchangeability', *Giorn. Ist. It. Attuari*, **27** (1964) 174–196. Translated in chapter 10 of de Finetti (1974).

Rudolf Carnap, *The Logical Foundations of Probability*, University of Chicago Press (1950, 1964).

L.J. Cohen, in *Behavioral and Brain Sciences* **4** (1981) 317–331.

L. Crisma, 'Sulla proseguibilita di processi scambiabili', *Rend. Matem. Trieste* **3** (1971) 96–124.

Donald Davidson, *Essays on Actions and Events*, Oxford, Clarendon Press (1980)

Bruno de Finetti, *Probabilismo: Saggio critico sulla teoria delle probabilità e sul valora della scienze*, Biblioteca di Filosofia diretta da Antinio Aliotta, Naples, Pirella (1931).

_____, 'La prevision: ses Lois Logiques ses sources subjectives', *Ann. Inst. H. Poincar* (Paris), **7** (1937) 1–68. Translated in Kyburg and Smokler (1964).

_____, 'Sulla proseguibilit di processi aleatori scambiabili', *Rend. Matem. Trieste*, **1**, 5 (1969) 3–67.

_____, 'La probabilita e la statistica nei rapporti con l'induzione, secondo i diversi punti di vista'. In *Corso C.I.M.E. su Induzione e Statistica.* Rome: Cremones. (1959). Translated as chapter 9 of de Finetti (1974).

_____, *Probability, Induction and Statistics*, New York, Wiley (1974).

_____, *Teoria delle Probabilità*, vol. 1, Einaudi (1970), *Theory of Probability*, New York, Wiley (1975).
_____, *Theory of Probability*, vol. 2, New York, Wiley (1975) 211–224.
_____, 'Probability: Beware of falsifications'! *Scientia* **111** (1976) 283–303.
_____, 'La Prevision...', *Annales de l'Institut Henri Poincar* **7** (1937). Translated in Kyburg and Smokler (eds.) (1980).
_____, 'Sur la condition d'equivalence partielle' (Colloque Geneve, 1937). *Act. Sci. Ind.* **739** (1938) 5–18, Paris, Herman. Translated in R. Jeffrey (ed.) (1980).
_____, 'Foresight,' *Studies in Subjective Probability*, Henry Kyburg and Howard Smokler (eds.), Huntington, N.Y., Krieger (1980).
Morris DeGroot and Stephen Feinberg, 'Assessing Probability Assessors: Calibration and Refinement,' in Shanti S. Gupta and James O. Berger (eds.), *Statistical Decision Theory and Related Topics III*, vol. 1, New York, Academic Press (1982) 291–314.
Persi Diaconis, 'Finite forms of de Finetti's theorem on exchangeability', *Synthese* **36** (1977) 271–281.
Persi Diaconis and David Freedman, 'Finite exchangeable sequences', *Ann. Prob.* **8** (1978a) 745–764.
_____, 'De Finetti's theorem for Markov chains', *Ann. Prob.* **8** (1978b) 115–130.
_____, 'Partial exchangeability and sufficiency', *Prof. Indian Stat. Inst. Golden Jubilee Int. Conf.*, Applications and New Directions (1978c) 205–236, J. K. Ghosh and J. Roy, eds., Indian Stat. Inst., Calcutta.
_____, *Studies in Inductive Logic and Probability*, vol. 2, Richard Jeffrey (ed.), Berkeley, University of California Press, 1980).
Persi Diaconis and Sandy Zabell, 'Some alternatives to Bayes's rule' in *Information Pooling and Group Decision Making*, Bernard Grofman and Guillermo Owen (eds.), Greenwich, Conn., and London, England, JAI Press, 25–38.
_____, 'Updating subjective probability' by the same authors, *Journal of the American Statistical Association* **77** (1982) 822–830.
Jon Dorling, 'Bayesian personalism, the methodology of research programmes, and Duhem's problem', *Studies in History and Philosophy of Science* **10** (1979) 177–187.
_____, 'Further illustrations of the Bayesian solution of Duhem's problem' 29 pp., photocopied (1982).
John Earman, *Bayes or Bust?* Cambridge, Mass.: MIT Press (1992).
David M. Eddy, 'Probabilistic reasoning in clinical medicine', in Kahneman Slovic, and Tversky (eds.) (1982).
Maurits Cornelius Escher, 'Ascending and Descending' (lithography, 1960), based on Penrose and Penrose (1958).
_____, *Escher on Escher*, New York, Abrams (1989).
William Feller, *An Introduction to Probability Theory and Its Applications*, vol. 1, 2nd ed. New York, Wiley (1957).
T. Ferguson, 'Prior distributions on spaces of probability measures', *Ann. Stat.* **2** (1974) 615–629.
James H. Fetzer (ed.), *Probability and Causality*, Dordrecht, Reidel (1988).
Ronald Fisher, *Smoking: The Cancer Controversy*, Edinburgh and London, Oliver and Boyd (1959).

H. Föllmer. 'Phase transitions and Martin boundary', *Springer Lecture Notes in Mathematics* **465** (1974) 305–317. New York, Springer–Verlag.
David Freedman, 'Mixtures of Markov processes', *Ann. Math. Stat.* **33** (1962a) 114–118.
_____, 'Invariants under mixing which generalize de Finetti's theorem', *Ann. Math. Stat.* **33** (1962b) 916–923.
David Freedman, Robert Pisani, and Roger Purves, *Statistics*, New York, W. W. Norton (1978).
Haim Gaifman, 'A theory of higher–order probability', in Brian Skyrms and William Harper (eds.), *Causality, Chance, and Choice*, Dordrecht, Reidel (1988).
Daniel Garber, 'Old evidence and logical omniscience in Bayesian confirmation theory' *Testing Scientific Theories*, ed. John Earman, Minneapolis, University of Minnesota Press (1983).
_____, 'Bayesianism with a human face' *Probability and the Art of Judgment*, Richard Jeffrey (ed.), Cambridge University Press (1992).
Clark Glymour, *Theory and Evidence*, Princeton University Press (1980).
I.J. Good, *Probability and the Weighing of Evidence*, London (1950).
_____, 'A. M. Turing's Statistical Work in World War II' *Biometrika* **66** 393–396 Minneapolis, (1979).
_____, *Good Thinking: The Foundations of Probability and Its Applications*, University of Minnesota Press (1983).
Nelson Goodman, *Fact, Fiction and Forecast*, Cambridge, Mass., Harvard University Press, 4th ed. (1983).
Stephen Jay Gould, *Full House*, New York, sec. 4, (1996).
Alan Hájek, *Probabilities and Conditionals*, Ellery Eells and Brian Skyrms (eds.), Cambridge University Press (1994).
Ned Hall, *Probabilities and Conditionals*, Ellery Eells and Brian Skyrms (eds.), Cambridge University Press (1994).
D. Heath and W. Sudderth, 'De Finetti's theorem on exchangeable variables', *Am. Stat.* **30** (1976) 188–189.
Carl G. Hempel, 'Studies in the logic of confirmation', *Mind* **54** (1945), reprinted in *Aspects of Scientific Explanation*, New York, The Free Press (1965).
_____, *Foundations of Natural Science*, New York, Prentice-Hall (1966).
_____, *Selected Philosophical Essays*, Cambridge University Press (2000).
Andrew Hodges, *Alan Turing, the Enigma*, New York, Simon and Schuster (1983).
Douglas R. Hofstadter, 'Metamagical Themas', *Scientific American* **248** 5 (May 1983) 16–26.
Janina Hosiasson-Lindenbaum, 'On confirmation,' *The Journal of Symbolic Logic* **5** (1940) 133–148.
Colin Howson and Peter Urbach, *Scientific Reasoning: the Bayesian approach*, La Salle, Illinois, Open Court, 2nd ed. (1993).
Christiaan Huygens, *De Ratiociniis in Ludo Aleae* (1657) ('On Calculating in Games of Luck')
Richard Jeffrey (ed.), *Studies in Inductive Logic and Probability*, Berkeley and Los Angeles, University of California Press, vol. 2 (1980).
_____, 'Risk and human rationality', *The Monist* **70** (1987).

_____, 'How to Probabilize a Newcomb Problem' in Fetzer (ed.) (1988).

_____, *The Logic of Decision*, New York, McGraw-Hill (1965) 2nd revised ed., University of Chicago Press (1983) further revision of 2nd ed. (1990).

_____, *Probability and the Art of Judgment*, Cambridge University Press (1992) 59–64.

_____, 'Causality in the Logic of Decision', *Philosophical Topics* **21**, 1 (1993) 139–151. University of Arkansas Press.

_____, 'Decision Kinematics' in Arrow et al. (eds.) (1996) 3–19.

_____, Petrus Hispanus Lectures 2000: *After Logical Empiricism*, Sociedade Portuguesa de Filosofia, Edições Colibri, Lisbon (2002) (edited, translated, and introduced by Antnio Zilhão).

James M. Joyce, 'Levi on causal decision theory and the possibility of predicting one's own actions', *Philosophical Studies* **110** (2002) 69–102.

Daniel Kahneman, Paul Slovic, and Amos Tversky (eds.), *Judgment Under Uncertainty*, Cambridge University Press (1982).

D. Kendall, 'On finite and infinite sequences of exchangeable events', *Studia Sci. Math. Hung.* **2** (1967) 319–327.

A. Khinchin, *Mathematical Foundations of Statistical Mechanics*, New York, Dover (1949).

Henry Kyburg, Jr. and Howard Smokler, *Studies in Subjective Probability*, 2nd ed., Huntington, N.Y., Robert E. Krieger (1980).

O.E. Lanford, 'Time evolutions of large classical systems, Theory and Applications', *Springer Lecture Notes in Physics* **38** (1975). New York, Springer-Verlag.

S.L. Lauritzen, 'On the interrelationships among sufficiency, total sufficiency, and some related concepts', Institute of Math. Stat., University of Copenhagen (1973).

_____, 'General exponential models for discrete observations', *Scand. I. Statist.* **2** (1975) 23–33.

Steven Leeds, 'Chance, realism, quantum mechanics', *J. Phil* **81** (1984) 567–578.

C.I. Lewis, *An Analysis of Knowledge and Valuation*, LaSalle, Illinois, Open Court (1947).

David Lewis, 'Probabilities of Conditionals and Conditional Probabilities' (1976), reprinted in his *Philosophical Papers, II*, Oxford University Press (1986).

Godehard Link, 'Representation Theorems of the de Finetti Type for (Partially) Symmetric Probability Measures,' *Studies in Inductive Logic and Probability*, vol. II, Richard C. Jeffrey (ed.), Berkeley and Los Angeles, University of California Press (1980) 208–231.

George W. Mackey, *Mathematical Foundations of Quantum Mechanics*, New York, W. A. Benjamin (1963).

P. Martin-Löf, *Statistiska Modeller*, Antechningar fran seminarer lsaret, 1969–70, uturbetade of R. Sundberg, Institutionen for Matematisk Statistik, Stockholms Universitet (1970).

_____, mimeographed lecture notes (1970).

_____, 'Repetitive structures and the relation between canonical and microcanonical distributions in statistics and statistical mechanics'. In Proc. Conference on Foundational Questions in Statistical Inference,

Department of Theoretical Statistics, University of Aarhus–Barndorf Nielsen, ed. P. Bluesild and G. Schon (1973).

_____, 'The notion of redundancy and its use as a quantitative measure of the discrepancy between a statistical hypothesis and a set of observational data,' *Scandinavian J. Statist.* **I**, I (1974) 3–18.

Robert Nozick, *The Normative Theory of Individual Choice*, Princeton University, Ph.D. dissertation (1963).

_____, 'Newcomb's problem and two principles of choice', in Rescher (ed.) (1969).

_____, Photocopy of Nozick (1963) with new preface; New York, Garland (1990).

Abraham Pais, *Subtle is the Lord..., the Science and Life of Albert Einstein*, Oxford University Press (1982).

L.S. Penrose and R. Penrose, 'Impossible objects: a special type of visual illusion', *The British Journal of Psychology* **49** (1958) 31–33.

C. Preston, 'Random Fields', *Springer Lecture Notes in Mathematics* **534** (1976). New York, Springer-Verlag.

George Polya, 'Heuristic reasoning and the theory of probability', *American Mathematical Monthly* **48** (1941) 450–465.

_____, *Patterns of Plausible Inference*, 2nd ed., Princeton University Press, vol. 2 (1968).

Hilary Putnam, *Mathematics, Matter and Method*, Cambridge University Press (1975) 293–304.

Wlodek Rabinowicz, 'Does practical deliberation crowd out self-prediction?', *Erkenntnis* **57** (2002) 91–122.

Frank Ramsey, 'Truth and Probability', *Philosophical Papers*, D. H. Mellor (ed.), Cambridge University Press (1990).

Michael Redhead, 'A Bayesian reconstruction of the methodology of scientific research programmes,' *Studies in History and Philosophy of Science* **11** (1980) 341–347.

Nicholas Rescher (ed.), *Essays in Honor of Carl G. Hempel*, Dordrecht, Reidel (1969).

Dov Samet, 'Bayesianism without Learning,' *Research in Economics*, **53** (1999) 227–242.

L.J. Savage, *The Foundations of Statistics*, New York, John Wiley and Sons (1954).

Schwartz, Wolfe, and Pauker, 'Pathology and Probabilities: a new approach to interpreting and reporting biopsies', *New England Journal of Medicine* **305** (1981) 917–923.

Abner Shimony, Personal communication, 12 Sept. (2002).

Brian Skyrms, *Causal Necessity*, New Haven, Conn., Yale University Press (1980).

_____, 'Dynamic Coherence and Probability Kinematics', *Philosophy of Science* **54** (1987) 1–20.

_____, *Choice and Chance: An Introduction to Inductive Logic*, Belmont, Calif., Wadsworth, 4th revised ed. (2000).

Paul Teller, 'Conditionalization and observation', *Synthese* **26** (1973) 218–258.

T. Tjur, 'Conditional probability distributions', Lecture Notes 2, Institute of Mathematical Statistics, University of Copenhagen (1974).

John Venn, *The Logic of Chance*, 3rd. ed. (1888), Chelsea Pub. Co. reprint (1962).

John von Neumann and Oskar Morgenstern, *Theory of Games and Economic Behavior*, Princeton University Press (1943, 1947).

Carl Wagner, 'Old evidence and new explanation I' in *Philosophy of Science* **64**, 3 (1997) 677–691.

_____, 'Old evidence and new explanation II' in *Philosophy of Science* **66** (1999) 283–288.

_____, 'Old evidence and new explanation III', *PSA 2000* (J. A. Barrett and J. McK. Alexander, eds.), part 1, pp. S165–S175, Proceedings of the 2000 biennial meeting of the Philosophy of Science Association, supplement to *Philosophy of Science* **68** (2001).

_____, 'Probability Kinematics and Commutativity', *Philosophy of Science* **69** (2002) 266–278.

Peter Whittle, *Probability via Expectation*, Heidelberg, Springer, 4th ed. (2000).

Sandy Zabell, 'The Rule of Succession,' *Erkenntnis* **31** (1989) 283–321.

Index